ShipPlotter

A User Guide for the Ship Spotting Enthusiast

by Lionel K Anderson MSc

ShipPlotter
A User Guide
for the Ship Spotting Enthusiast

First Edition (greyscale) 2009 by Las Atalayas
Publishing

Author and Editor Lionel K Anderson MSc

ISBN 978-1-4452-1531-0

Design and Typesetting by kenandglen.com

Contents

1 Introduction	1
2 Automatic Identification System	3
3 Radios for ShipPlotter	13
4 Software Installation	19
5 First Time Use	23
6 ShipPlotter Menus	31
7 ShipPlotter Toolbar	55
8 Charts and Calibration	69
9 Sharing	77
10 Google Maps & Web Pages	83
11 Support	97
Appendix A	100
Appendix B	102
Bibliography	118
Index	119

Disclaimer

Whilst the ShipPlotter software can be used by mariners, small boat owners and yachtsmen this book is written for and intended solely for the hobbyist and ship spotting enthusiast.

It's contents therefore should not be used as any sort of guidance or advice for those who are not firmly fixed to their seats in the comfort of their homes on dry ground!

Acknowledgements

I'd first like to thank René Harte for his photograph of the Celine that was used for the basis of the books cover.

I'd like to thank Aleksi Lindström, Bjørn Poulsen, Daniel Ferro, Dave van Spronsen, Hannes van Rijn, Holger Jaschob, Jan Melchers, K Mikkelsen, Ken Lubi, Wim van der Moolen for permission to include their excellent ship photographs which have been used to help illustrate this book.

I'd also like to thank Curt Deegan for permission to include part of his SP2GM, Ship9 and Google.html visual basic scripts.

And of course thanks to Bev Ewen-Smith for creating and maintaining the ShipPlotter software and for his never ending help and support to all its users.

About the Author

Lionel K Anderson spent his schooldays living on the Isle of Wight. His parents' house was just a hundred metres from the waters edge of the Spithead. From there it was possible for him to view the mass of transatlantic passenger shipping in the Solent moving to and from Southampton docks. Giant passenger liners such as the Queen Mary, Queen Elizabeth, France, Canberra and Oriana passed by regularly and often within just one mile of the shoreline. As well as the passenger traffic there was a large amount of cargo traffic bound to and from the Southampton Docks, oil tankers for the Fawley Refinery and of course naval ships of all sizes from small minesweepers to giant aircraft carriers for Portsmouth Naval Base.

Those days spent observing ship movements in the 1950's laid the seed for the interest he still has in watching ships although nowadays it is done from the warmer shores of eastern Spain.

A professional electronics engineer with a basic discipline of radar and microwave engineering although now retired he continues his interests and activities in electronics and computing. Having thought of building a small marine radar surveillance to monitor local shipping in the Balearic Sea he was delighted to discover the Ship Plotter software program that now fully meets his needs in that area.

Glossary

AIS	Automatic Identification System
AIS-SART	AIS- Search and Rescue Transponder
AVI	Audio Video Interleave
BMP	Bitmap graphic format file extension
BSB	BSB a proprietary chart format
CLB	Calibration file extension
COAA	Centro de Observação Astronómica no Algarve
CSTDMA	Carrier Sense Time Division Multiple Access
DGPS	Differential Global Positioning System
DSC	Digital Selective Calling
ETA	Estimated Time of Arrival
FCS	Frame Check Sequence
FIR	Finite Impulse Response
FTP	File Transfer Protocol
GMDSS	Global Maritime Distress Safety System
GMSK	Gaussian Minimum Shift Keying
GPS	Global Positioning System
GPX	GPS Exchange Format
HDLC	High-Level Data Link Control
HTTP	HyperText Transfer Protocol
I/O	Input/Output
IMO	International Maritime Organisation
JPEG	Joint Photographic Experts Group
JPG	JPEG graphic format file extension
KML	Keyhole Markup Language
MMSI	Maritime Mobile Service Identity
NIMA	National Imagery and Mapping Agency
NMEA	National Marine Electronics Association
NTP	Network Sharing Protocol
OSM	Open Street Map
PC	Personal Computer
SOLAS	International Convention For The Safety Of Life At Sea
SAR	Search and Rescue
SOTDMA	Self-Organizing Time Division Multiple Access
TCP/IP	Transmission Control Protocol/Internet Protocol
TDMA	Time Division Multiple Access
UDP/IP	User Datagram Protocol/Internet Protocol
USB	Universal Serial Bus
UTC	Coordinated Universal Time
VEC	A graphic file extension associated with Daylon Vectre Vector Graphic
VHF	Very High Frequency
VMAP	Vector Smart Map
XML	Extensible Markup Language

Tug - Fairplay I
Photo Courtesy and Copyright of Aleksi Lindström

Chapter 1

Introduction

ShipPlotter is a unique piece of software that enables a user to have a live radar type display of shipping in their local coastal region or other regions and waterways around the world. The software decodes radio signals, received using a VHF radio receiver or scanner, from ships transmitting digital data using the marine Automatic Identification System (AIS).

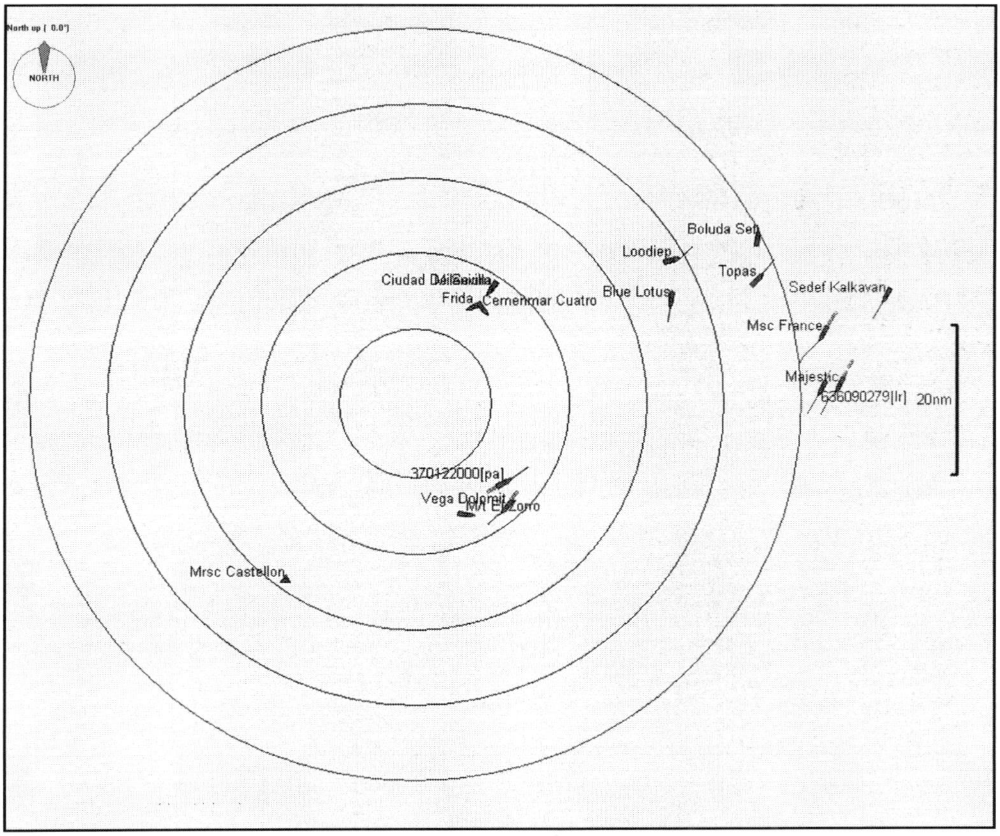

Figure 1 - ShipPlotter Radar Type Display

The decoded data includes the following information

- Ships' MMSI (Maritime Mobile Service Identity)
- Ships' Name
- Ships' Call Sign
- Ships' IMO (International Maritime Organization) Number
- Ships' Country of Registration
- Ships' Destination
- Ships' ETA (Estimated Time of Arrival) at it's Destination
- Ships' Current Position (Latitude & Longitude)
- Ships' Heading
- Ships' Course over the Ground
- Ships' Speed over the Ground
- Ships' Dimensions (Beam & Draught)
- Ships' Bearing from the Receiving Station
- Ships' Range from the Receiving Station
- Ships' Navigational Status

ShipPlotter visually displays the position and identification of each ship either as a radar view on a circular display centred on the user's position or on a chart created from a graphic image file, a satellite image download or a downloaded Open Street Map (OSM).

ShipPlotter also displays the decoded messages as a window of plain text on the PCs' screen as well as a tabular list of ships with their decoded information.

As well as displaying the position of ships received using a local radio receiver ShipPlotter can display data shared by other ShipPlotter users by connecting to the Internet. Thus a user in the centre of Europe can see quasi-live shipping movements from all around the globe.

Other software enabling ship positions and movements to be displayed over the Internet using standard Web pages can access the ShipPlotter data files. Visual interfaces using Google Maps and even Google Earth server can be easily implemented.

As well as displaying ship movements on the seas and oceans of the world ShipPlotter can also display movements on some inland waterways such as the River Rhine, the Panama Canal or St Lawrence Seaway.

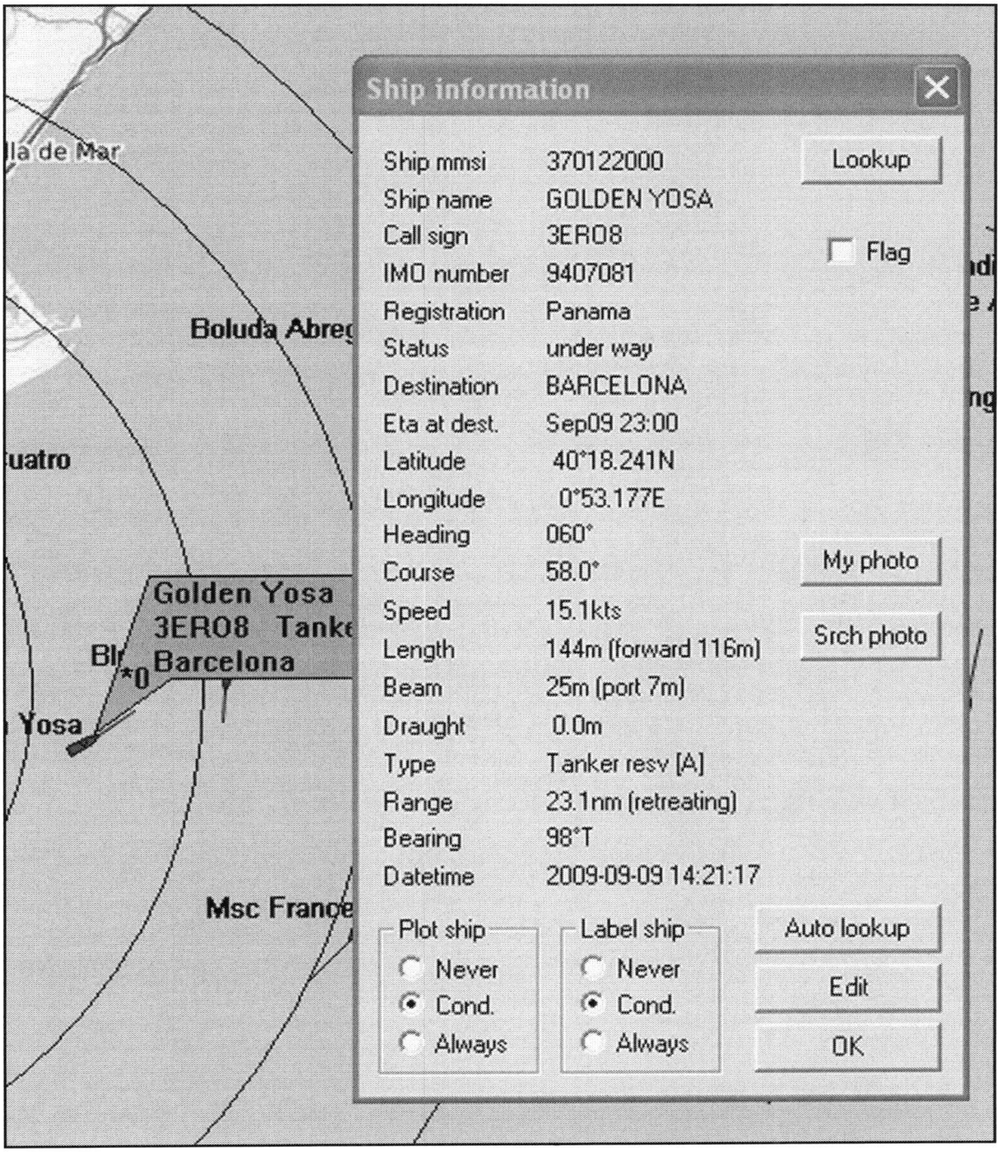

Figure 2 - A pop-up window of plain text showing decoded message for one ship

Cargo (Cement Carrier) - Cemenmar Cuatro
Photo Courtesy and Copyright of Wim van der Moolen

Chapter 2

Automatic Identification System

The Shipborne Automatic Identification System (AIS) is a system used by ships mainly for the identification of vessels at sea. It is a ship and shore based broadcast system, operating in the VHF maritime band using the two frequencies 161.975 and 162.025 MHz. It is capable of sending and receiving ship information such as identity, position, course, speed, ship particulars and cargo information to and from other ships as well as suitably equipped aircraft and shore stations. The system can handle over 2,000 reports per minute and provides updated information as often as every two seconds. AIS helps to overcome the difficulty of identifying ships that are not in sight of each other such as in the darkness of night, in foggy conditions, or those without radar or in radar blind arcs and of course those over the horizon.

The International Maritime Organisations' (IMO) Safety of Life at Sea (SOLAS) requires AIS to be fitted aboard all ships greater than 300 tons.

On board ship AIS will be used in navigation primarily for collision avoidance but because of the technical limitations of VHF radio communications and because not all vessels will be equipped with AIS the system is more regularly used as a means of lookout and to determine the risk of collision rather than as an automated collision avoidance system.

When a ship is navigating at sea the positions, movements and identities of other ships in the vicinity are obviously important to navigators to enable them to make decisions thus avoiding collision with other ships or other dangers. In the past whilst visual and audio watch keeping were primarily used for this purpose a lack of positive identification of the targets on the displays could sometimes prevent action in time to avoid collision.

Although AIS provides just a very basic graphical display its data can be integrated with other Navaids such as Radar and Electronic Chart Display and Information Systems (ECDIS) thus providing more comprehensive navigational information on a single display.

AIS is also key in co-ordinating marine search and rescue operation resources as it is important to know the position and navigation status of ships in the vicinity of the ship or person in distress. AIS provides a much better picture of the resources for on-scene operation even though the range of AIS is limited to the VHF radio range that is normally about 30 to 50 nautical miles.

AIS transponders send the following data as often as every two minutes:

MMSI number of vessel - vessel's unique identification
Navigation status - e.g. at anchor, under way using engine, not under command etc
Rate of turn - right or left, 0 to 720 degrees per minute
Speed over ground - 0.1 knot resolution from 0 to 102 knots
Longitude - to 1/10000 minute and Latitude - to 1/10000 minute
Course over ground - relative to true north to 0.1 degree
True Heading - 0 to 359 degrees from a gyrocompass
Time stamp - UTC time accurate to nearest second.

In addition, the following data is broadcast every 6 minutes:

MMSI number - vessel's unique identification
IMO number - this number always remains with the ship
Radio call sign - international radio call sign assigned to vessel
Name - Name of vessel
Type of ship/cargo
Dimensions of ship - to nearest meter
Location of positioning system's (e.g. GPS) antenna onboard the vessel
Type of positioning system - usually GPS or DGPS
Draught of ship - 0.1 meter to 25.5 meters
Destination - max 20 characters
ETA at destination - UTC month/date hour:minute

Class A AIS

There are two types of shipboard AIS available: Class A for larger ships and for those vessels subject to the Safety of Life at Sea Convention using

AIS Class A	Class A has been mandated by the International Maritime Organization (IMO) for vessels of 300 gross tonnage and upwards engaged on international voyages, cargo ships of 500 gross tonnage and upwards not engaged on international voyages, as well as passenger ships (more than 12 passengers), irrespective of size.
AIS Class B	Class B provides limited functionality and is intended for non-SOLAS vessels. It is not mandated by the International Maritime Organization (IMO) and has been developed for non-SOLAS commercial and recreational vessels.
AIS Base Station	Base Stations are provided by an aids to navigation authorities to enable the ship to shore / shore to ship transmission of information. Networked AIS Base Stations can assist in providing overall maritime domain awareness.
AIS aids to navigation (AtoN)	AIS AtoN provide an opportunity to transmit position and status of buoys and lights through the same VDL, which can then show up on an electronic chart, computer display or compatible radar.
AIS SART	Search and Rescue Transmitters using AIS can be used to assist in determining the location of a vessel in distress.
AIS on Search and Rescue (SAR) Aircraft	Search and Rescue Aircraft may use AIS to assist in their operations.

Figure 3 - Types of AIS
(Australian Maritime Safety Authority)

Self Organizing Time Division Multiple Access (SOTDMA) technology and Class B normally used for the small commercial boat, fishing boat and yachting using Carrier-Sense Time Division Multiple Access (CSTDMA) technology.

Each Class A AIS system consists of a 12.5W VHF transmitter, an integral global navigation satellite system receiver, two VHF Time Division Multiple Access (TDMA) receivers, one VHF Digital Selective Calling (DSC) receiver, and standard marine electronic data interface to shipboard display and sensor systems (AIS Schematic). Position is normally derived from an external navigation receiver and timing information derived from the integral GPS receiver, although the integral receiver does provide position information as a backup. Precise position information in coastal and inland waters is available too, from either a dGPS receiver or equivalent information broadcast by an AIS base station. Other information broadcast by the AIS, if available, is electronically obtained from shipboard equipment through standard marine data connections.

All AIS-equipped ships will normally provide heading information and course and speed over ground. Other information, such as rate of turn, angle of heel, pitch and roll, and destination and ETA can also be provided.

Some typical decoded Class A data would look like this:

247629000;(9164782);(IBMK);CALA PEDRA;Cargo ship ;-> Sep05 08:00 VALENCIA;190 24 7.1 0 0; 090905 040045;serial#1(A)[5]
247629000;under way ;000°';21.9kt;40.151333N;0.534333E;221.0°;216°;39s; 090905 040041;serial#1(A)[1]

Class B AIS

Class B AIS generally consists of a 2W VHF transmitter, a GPS receiver and two VHF receivers, one of which is able to decode DSC transmissions as well as AIS.

The data transmitted by Class B AIS is nearly identical to the Class A, except that it:

- Has a reporting rate less than a Class A
- Does not transmit the vessel's IMO number
- Does not transmit ETA or destination
- Does not transmit navigational status
- Is only required to receive, not transmit, text safety messages
- Is only required to receive, not transmit, application identifiers (binary messages)
- Does not transmit rate of turn information
- Does not transmit maximum present static draught

AIS Transponder

The AIS transponder normally works in an autonomous and continuous mode, regardless of whether it is operating in the open seas or coastal or inland areas. Transmissions use 9.6kb Gaussian Minimum Shift Keying (GMSK) FM modulation using High-level Data Link Control (HDLC) packet protocols. Although only one radio channel is necessary, each station transmits and receives over two radio channels to avoid interference problems, and to allow channels to be shifted without communications loss from other ships. The system provides for automatic contention resolution between itself and other stations, and communications integrity is maintained even in overload situations. Class A AIS can tune over the whole 156.025 -162.025 MHz VHF maritime band, while Class B AIS is limited to 161.5 - 162.025 MHz.

Each Class A AIS station determines its own transmission schedule or slot based upon data link traffic history and a knowledge of future actions by other stations. Class B AIS is a "polite" listen-before-transmitting system that will transmit on the first available slot. A position report from one AIS station fits into one of 2250 time slots established every 60 seconds. AIS stations continuously synchronize themselves to each other, to avoid overlap of slot transmissions. Slot selection by a Class A AIS station is randomised within a defined interval, and tagged with a random timeout of between 0 and 8 frames. When a station changes its slot assignment it pre-announces both the new location and the timeout for that location. In this way those vessels will always receive new stations, including those stations that suddenly come within radio range close to other vessels.

The required ship reporting capacity according to the IMO performance standard amounts to a minimum of 2000 time slots per minute, though the system provides 4500 time slots per minute. The SOTDMA broadcast mode

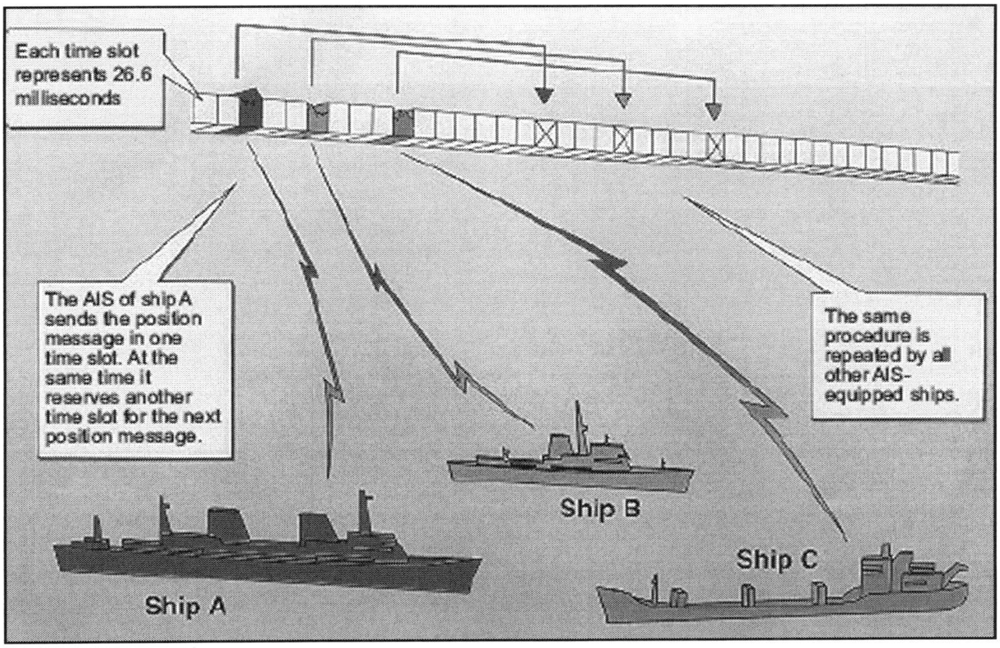

Figure 4 - Self Organizing Time Division Multiple Access (SOTDMA)
(Australian Maritime Safety Authority)

allows the system to be overloaded by 400 to 500% through sharing of slots, and still provide nearly 100% throughput for ships closer than 8 to 10 NM to each other in a ship-to-ship mode. In the event of system overload, only targets further away will be subject to dropout in order to give preference to nearer targets that are a primary concern to ship operators. In practice, the capacity of the system is nearly unlimited, allowing for a great number of ships to be accommodated at the same time.

The system coverage range is similar to other VHF applications, essentially depending on the height of the antenna. Its propagation is slightly better than that of radar, due to the longer wavelength, so it's possible to “see” around bends and behind islands if the landmasses are not too high. A typical value to be expected at sea is nominally 20 nautical miles. With the help of repeater stations, the coverage for both ship and VTS stations can be improved considerably.

Other types of AIS include base stations, aids to navigation stations, search and rescue transmitters (AIS-SART) and stations used on search and rescue aircraft.

AIS Message Types

There are many different AIS message formats. The commonest is the Type 1 message that contains the MMSI number and the dynamic information about position, course and speed. Ships send these messages every few seconds if they are under way and at speed. No information about ship name, IMO number, destination, dimensions, etc. is sent in a Type 1 message.

The Type 5 messages contain the static information like the ship name, IMO number, dimensions, destination and ETA and are only sent every 6 minutes. Listening on a single channel means the static message will be received once every 12 minutes because they alternate on the two VHF channels. Interference or a weak signal means that a Type 5 message may be missed and it might take 24 minutes to collect the static information.

The table below shows the individual messages that have been defined for AIS in the NMEA standard.

01 - Position Report with SOTDMA
02 - Position Report with SOTDMA
03 - Position Report with ITDMA
04 - Base Station Report
05 - Ship static and voyage related data
06 - Addressed Binary Message
07 - Binary Acknowledge
08 - Binary Broadcast Message
09 - Standard SAR Aircraft position report
10 - UTC and Date inquiry
11 - UTC and Date response
12 - Addressed safety related message
13 - Safety related Acknowledge
14 - Safety related Broadcast Message
15 - Interrogation
16 - Assigned Mode Command
17 - GNSS Binary Broadcast Message
18 - Standard Class B CS Position Report
19 - Extended Class B Equipment Position Report
20 - Data Link Management
21 - Aids-to-navigation Report
22 - Channel Management
23 - Group Assignment Command
24 - Class B CS Static Data Report

Fig 5 - AIS Message Types

Tanker (Crude Oil) - Siri Knutsen
Photo Courtesy and Copyright of Hannes van Rijn

Chapter 3
Radios for ShipPlotter

To receive the AIS information a VHF band radio receiver tuned to one of the AIS channels is required. There are a wide range of receivers readily available and amongst these are marine radios, Ham radios, scanners as well as special to purpose AIS receivers.

A VHF radio or scanner can be used as shown in Figure 5. Here the audio output of the receiver is connected to the Sound Card of a laptop or personal computer.

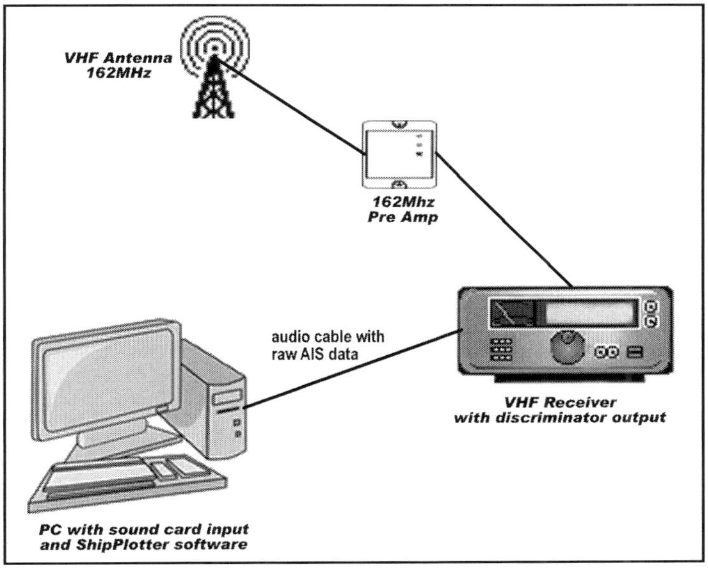

Fig 6 - ShipPlotter AIS installation with VHF radio receiver

The output from most scanners and VHF receivers is filtered and amplified to provide audio signals best for listening to with the human ear. AIS is a digital signal and the filtering and amplification may heavily distort the digital signal thus to get the best results it is better to use the unfiltered audio from a receiver. That signal can be found at the discriminator output of the receiver. The discriminator is the 'heart' of an FM receiver as it is the part of the receiver circuitry that converts the modulated radio frequency waves into the audio output.

As only a few scanners have a discriminator output as a standard feature then to decode and monitor digital signals to best effect it is essential to add a "discriminator tap" to the receiver circuitry. To do this then it is necessary to make a modification to the receivers' circuit board. Information on how to do this for a number of different scanners and VHF receivers can be found at Rene Harts' excellent website http://www.discriminator.nl/index-en.html. On that site can be found pictures and descriptions of almost 150 scanners, receivers, pagers and (marine) VHF transceivers that have been modified with a discriminator tap.

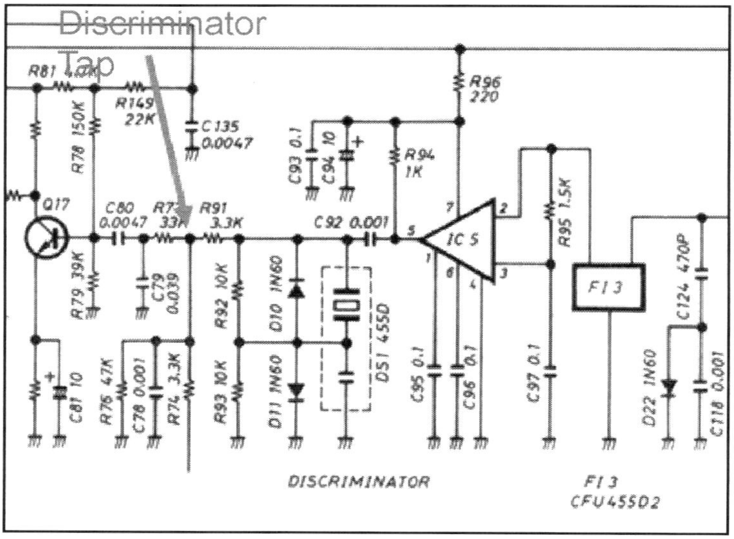

Fig 7 - Part of the circuit of an Icom IC-M80 marine receiver showing the discriminator output point (http://www.discriminator.nl/ic-m80/index-en.html)

The alternative to using a standard VHF receiver or scanner is to use a receiver especially designed for AIS. Here the receiver decodes the received data stream and sends a stream of NMEA (National Marine Electronics Association) format serial data to the PC as shown in Figure 8.

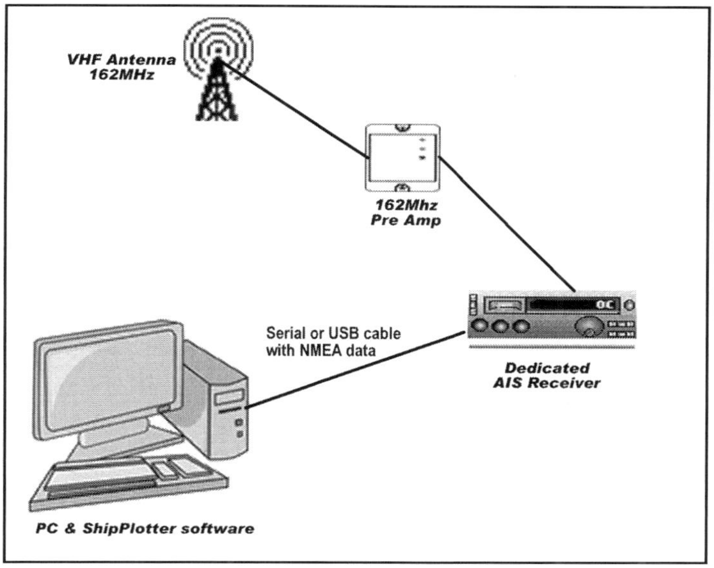

Fig 8 - ShipPlotter AIS installation with dedicated AIS radio receiver

Two examples of such a receiver are the Smart Radio SR162 and the Comar AIS-2-USB. The Comar AIS-2-USB receiver is a low cost Dual Channel Parallel AIS receiver specifically designed for use with a Laptop or PC and combines two AIS receiver monitoring both AIS channels simultaneously into one box. Unlike many AIS receivers that claim to be dual channel the AIS-2-USB is a true parallel AIS receiver, ensuring the most up to date positions and information of all vessels in the vicinity are received and decoded. The AIS-2-USB does not require a separate PSU as it is powered via the USB port.

Fig 9 - Comar AIS-2-USB

Fig 10 - SR162 AIS Receiver

The Comar AIS-2-USB Receiver can be purchased in the UK and Europe from JGTechnologies in Weymouth (http://www.jgtech.com).

Fig 11 - Garex MP-3 Pre Amp

A VHF Pre-Amp fitted close to the antenna can boost the received signal and increase the range and number of ships data received. It is important that such an amplifier be designed for use in the maritime VHF range of 156 - 162MHz as that should improve the readability of weak signals. A good preamplifier will also reject out-of-band signals thus further improving the quality of the in-band signals.

An excellent model is the Garex MP-3 Marine Band Preamplifier. The manufacturers supply an AIS model where the amplifier is peak tuned to 162MHz. The AIS MP-3 can be purchased in the UK and Europe from Garex Electronics (http://www.garex.co.uk).

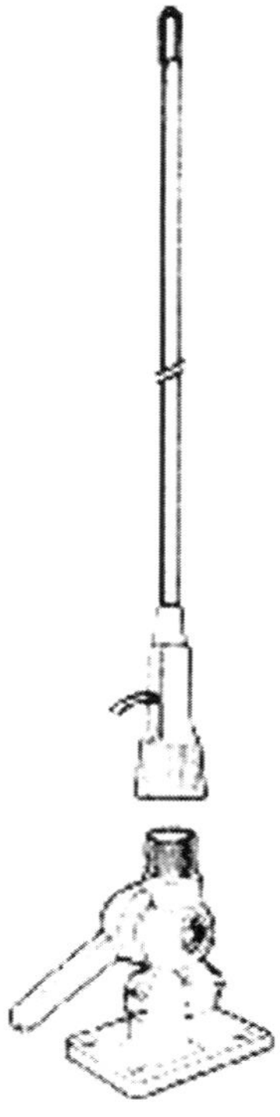

Fig 12 - Antenna

A good and well-sited antenna is essential as it is the first point in the system at which the AIS radio signal is received. Any marine VHF antenna should perform well but good results can also be obtained from Ham antennas designed for the 144Mhz/2 metre band.

For the hobbyist looking to make his or her own low cost antenna then the J-Pole design is well worth looking at as these can be constructed economically out of ½ inch copper pipe. A fine article on how to do it can be found at http://www.dxzone.com/cgi-bin/dir/jump2.cgi?ID=14332.

Figure 13 shows the basic design and how the component dimensions may be calculated. For an AIS 162Mhz J-Pole they are as follows:

A = 132.6cm B = 43.9cm C =4.3cm D = 4.3cm

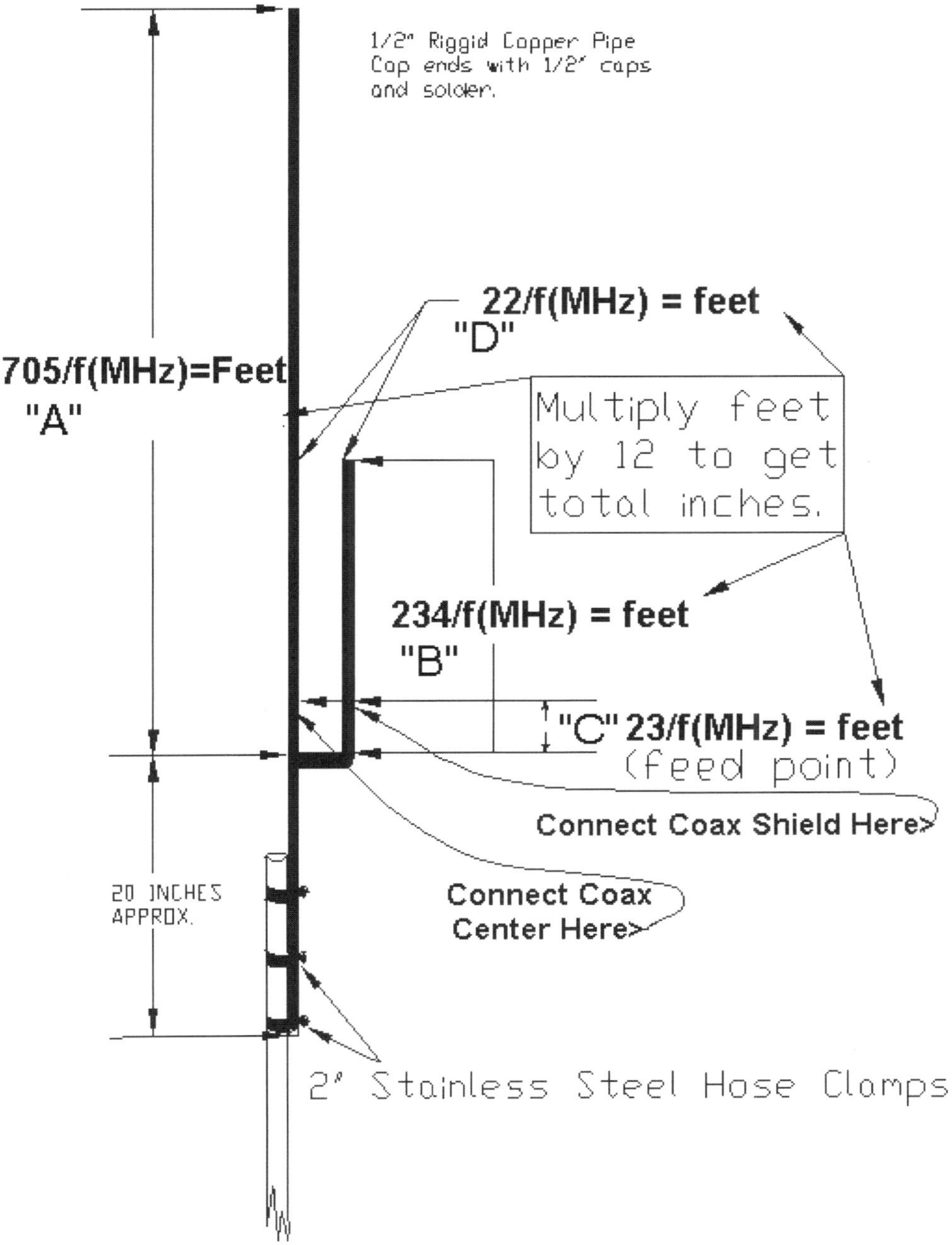

Fig 13 - J-Pole Antenna Calculator
(http://www.dxzone.com/cgi-bin/dir/jump2.cgi?ID=9284)

Search And Rescue Vessel - Salvamar Algeciras
Photo Courtesy and Copyright of Ken Anderson

Chapter 4

Software Installation

ShipPlotter has been designed to work on the Microsoft Windows platform and should run without difficulty on any modern PC or Laptop. Depending on the type of radio receiver it may also be necessary to have a compatible sound card installed in the computer. A minimum of a Pentium type processor is recommended as the real time processing of the AIS messages is processor intensive and requires a machine of reasonable performance. A minimum of 2Mb RAM is recommended to display large charts.

To receive the AIS messages a dedicated AIS receiver with a serial output is recommended. Alternatively a scanner or a general-purpose VHF band radio receiver tuned to one of the AIS channels may be used.

Download and Install the ShipPlotter Software

Before downloading the software it is recommended that there is a folder on the PC to keep any software downloaded to the computer. A good idea is to name this folder DOWNLOADS or ZIP as many downloaded files are commonly compressed in the Zip format.

It is also recommended that a new folder be created on the main computer hard drive for the ShipPlotter program and files. e.g. C:\ShipPlotter.

The ShipPlotter software is readily available and can be downloaded free of charge from http://www.coaa.co.uk/shipplotter.htm. Different language versions are available in English, French Dutch and Norwegian.

Having selected the file to download a dialog box, Figure 14 over page, will be displayed. Click on the Save button and then enter "C:\Zip" or "C:\Downloads" as the folder where the compressed ShipPlotter file will be stored as shown in Figure 15 over page.

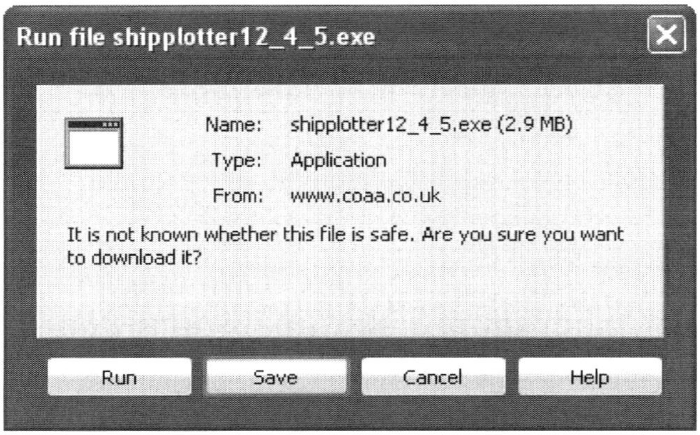

Figure 14 - Download dialog

Once the file has been safely downloaded close the Internet connection and then open the folder where the downloaded file has been stored.

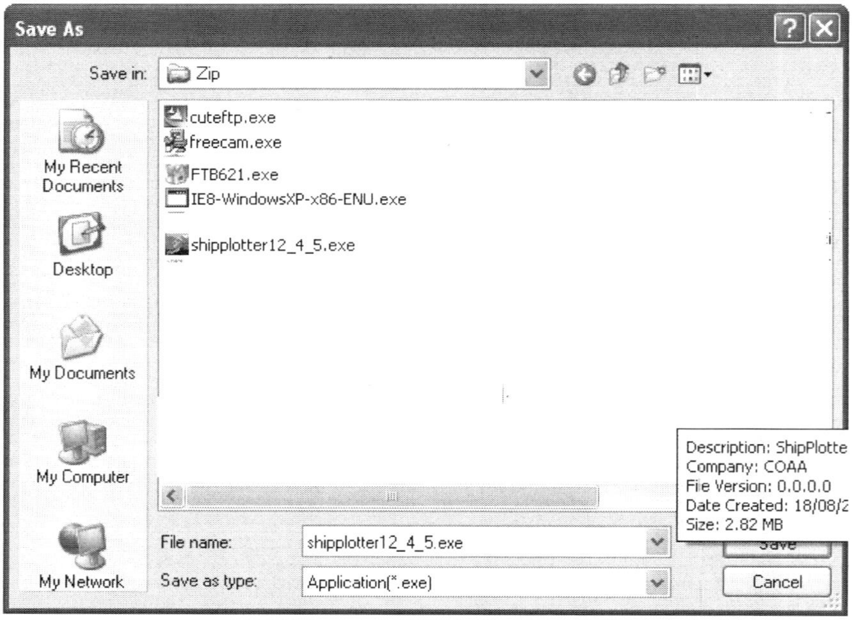

Figure 15 - Download destination

Next select the folder where the ShipPlotter program and files will be installed. It is recommended that user does **not accept** the default folder but ensures that "**C:\Shipplotter**" is specified as the target destination folder as shown in Figure 16 on the next page.

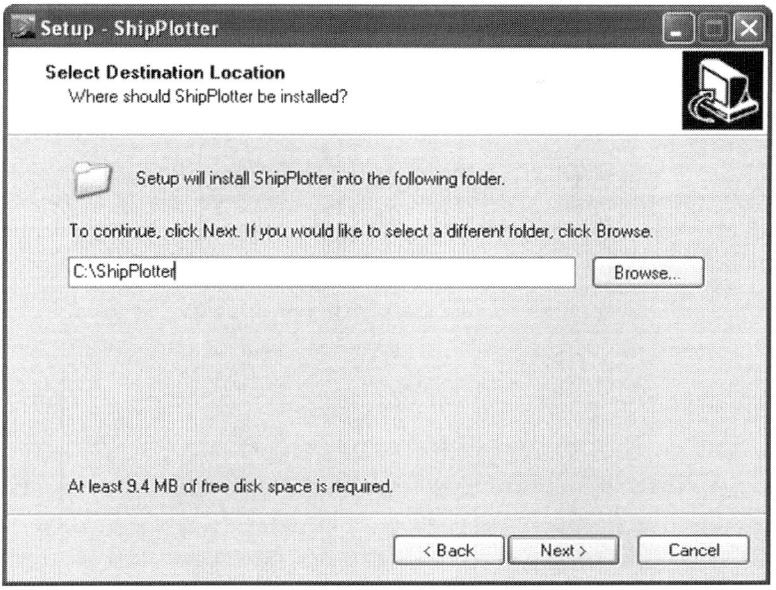

Figure 16 - Installation destination

When the software has been installed and before running the ShipPlotter program it is necessary to move some folders that have been created in the "*C:\Documents and Settings\Administrator\Application Data\COAA\ShipPlotter*" folder to the main "*C:\ShipPlotter*" folder.

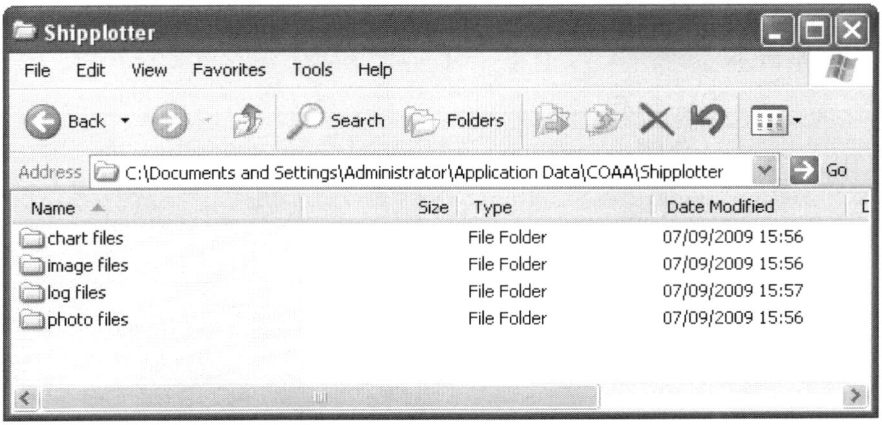

Figure 17 - Move these four folders to the C:\Shipplotter folder

*N.B. If the software was installed with the user logged in, as other than 'Administrator' then the folders to be moved will be found in the C:\Documents and Settings**USERNAME**\Application Data\COAA\ShipPlotter folder.*

Container Ship - MSC Eloise
Photo Courtesy and Copyright of Dave van Spronsen

Chapter 5

First Time Use

Having installed the software then on launching the ShipPlotter program for the first time the user will be prompted to register the software. The software once downloaded can be used freely for 21 days. After that time it must be registered. Registration can be done swiftly and securely on-line and costs only Euro €25 for personal use (VAT is applicable for customers within the EU). Note that COAA who author and provide the program strongly recommend that users ensure that ShipPlotter performs to their complete satisfaction before registering.

The registration is done online and payment made securely by credit card. Once the registration has been processed an email containing the registration key will be sent to the user and all that is then required is for the user to enter that key into the Registration Box shown in Figure 18.

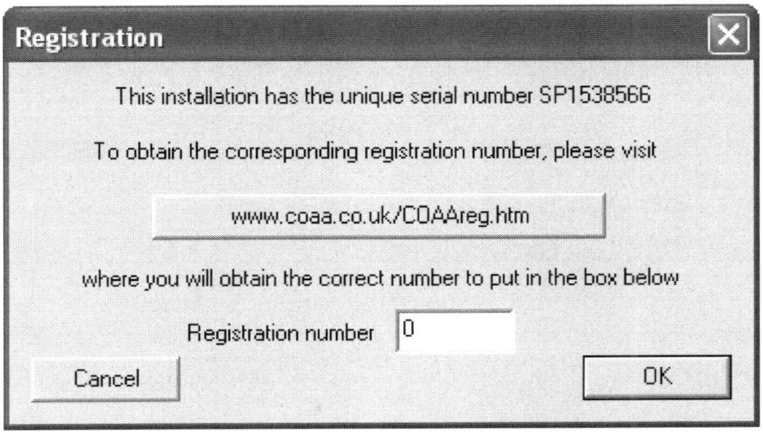

Figure 18 - ShipPlotter Software Registration

The first screen to then be displayed will probably be the 'Signal View' that shows the raw signal received by PC from the VHF Radio/Scanner, if used and connected. *N.B. If using an AIS Receiver like the Comar AIS-2-USB no signals will ever be displayed here, as the data will already be decoded.*

Selecting *"View, Chart"* from the Menu Bar or clicking on the "Chart Button" will result in a blank blue screen with the prompt to "Load a Chart before using this option".

Before loading a chart is highly recommended that the user enters the geographical latitude and longitude of the home station where the AIS receiver is located.

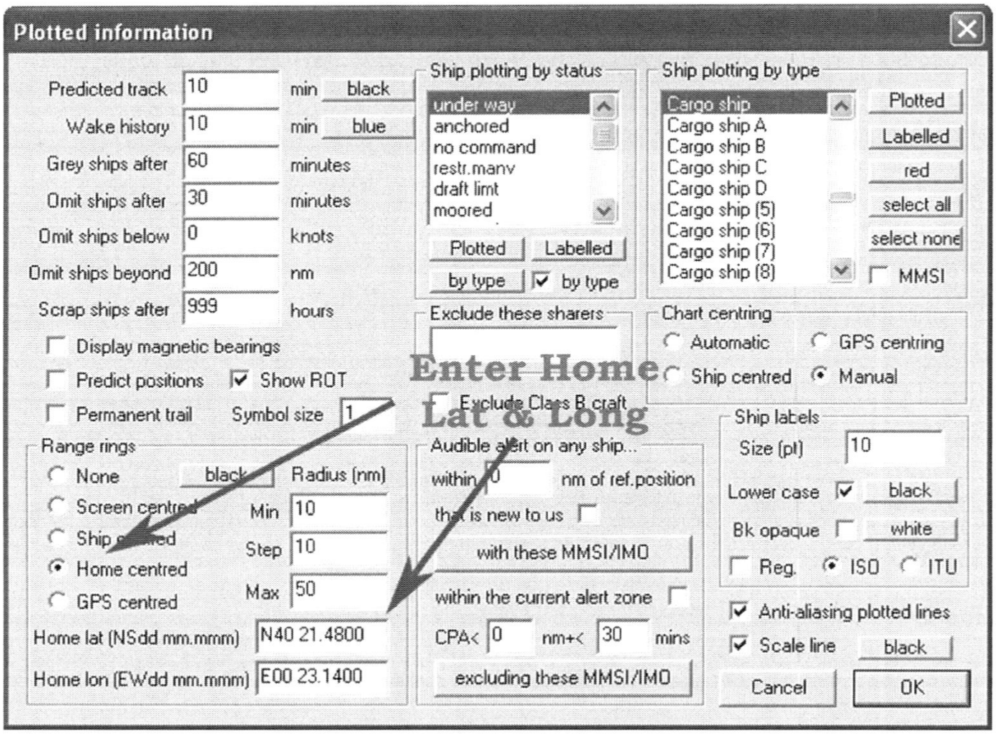

Figure 19 - Set Home Latitude & Longtitude Co-ordinates

To do this the user should click on the "Customize Button" that will display the form shown in Figure 19.

The user should then click on the Home Centred check box and enter the Latitude and Longitude co-ordinates for the home station. If the co-ordinates are not known then they can be determined using a useful "Lat-Long Finder" that can be found at http://www.satsig.net/maps/lat-long-finder.htm.

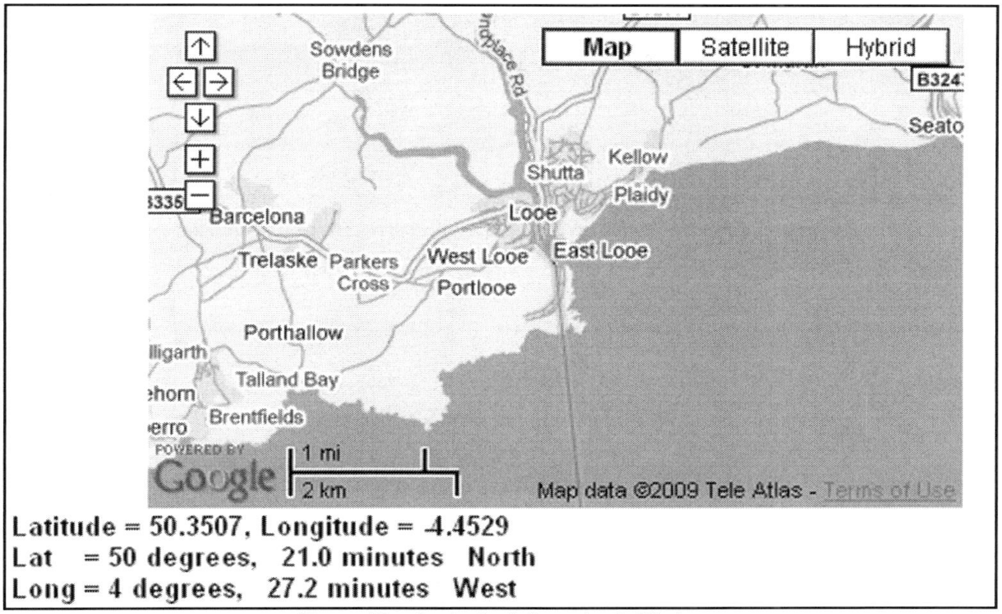

Figure 20 - Latitude & Longtitude Finder
(http://www.satsig.net/maps/lat-long-finder.htm)

A typical result is shown in Figure 20. This program can also be used when calibrating a chart as will be discussed later.

Charts

Before continuing further it is essential to have a chart region readily available of the area of the user wishes to monitor. The chart must be a graphic file in a standard format such as BMP or JPG.

There are a number of different ways of generating a chart. Using a flatbed scanner and some software to scan a hard copy chart, map or chart is one way. Here once a basic chart has been created it may be necessary to "tweak it" using a conventional graphics editor program such as Photoshop or PaintShopPro to resample the digitised chart to the minimum size in pixels that allows all the detail to be clearly legible, to rotate the chart so that it is the correct way up and normally with North at the top and possibly reduce the number of colours so that the image file is no larger than really required.

An alternative is to download a prepared chart from a web site such as http://www.selkieman.co.uk/coggie/charts or by using a chart generator such as that found at http://emit.demon.co.uk/map2.php. More information on this can be found in the ShipPlotter Help Tutorial.

Figure 21 shows a chart generated by the ShipPlotter map generator program. As can be seen the user only has to enter the size of the chart image required, the latitude and longitude of the AIS receiving station and the radius in nautical miles from the receiving station that the chart is required to cover. It is recommended that the radius be kept to a reasonable distance, better smaller than larger but it depends on location. With modern PCs an image size of 800x600 or 1024x768 might be better choices.

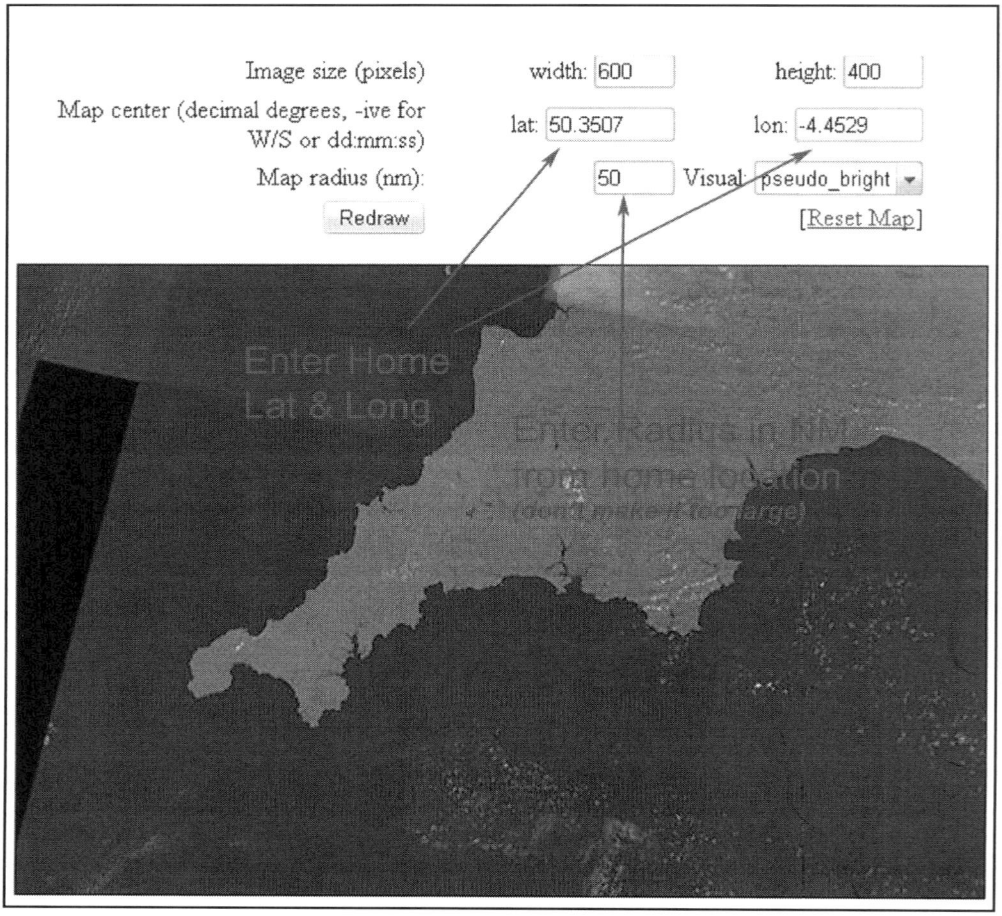

Figure 21 - ShipPlotter Chart Maker
(http://emit.demon.co.uk/map2.php)

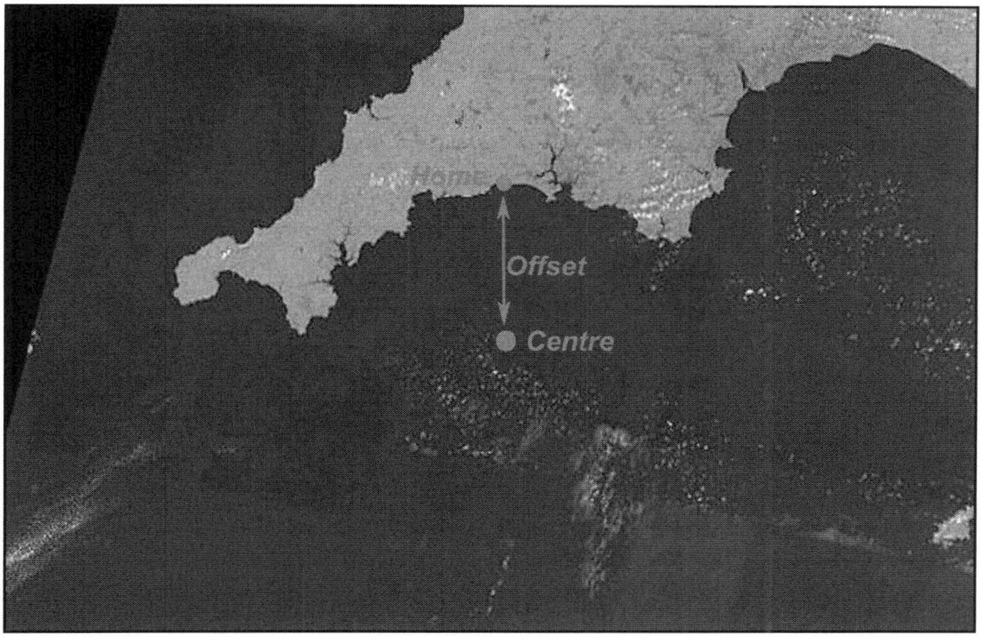

Figure 22 - Chart with centre shifted to reduce land mass
(http://emit.demon.co.uk/map2.php)

Once the chart has been generated it should be saved into the ShipPlotter Charts Folder (C:\ShipPlotter\Chart Files) along with the CLB calibration file that is generated by the program at the same as the chart file is created. e.g. Chart_1.jpg for the chart and Chart_1.clb for the calibration file.

Hint: As *charts produced in this way are centred on the entered home co-ordinates a large amount of the chart may be the landmass to one side of the home location. Obviously not many ships will be found on land! A good idea is to offset the co-ordinates used to generate the chart so that the chart is centred in the sea with only a reasonable amount of land mass plotted. Figure 22 shows the effect of offsetting the latitude by 0.5 deg.*

Basic Options

Having created the first chart some basic options must be set up before continuing any further.

From the ShipPlotter Menu Bar first select *"Options"* and then *"Directories"* as shown in Figure 23 over page. Then select each folder in turn Log files,

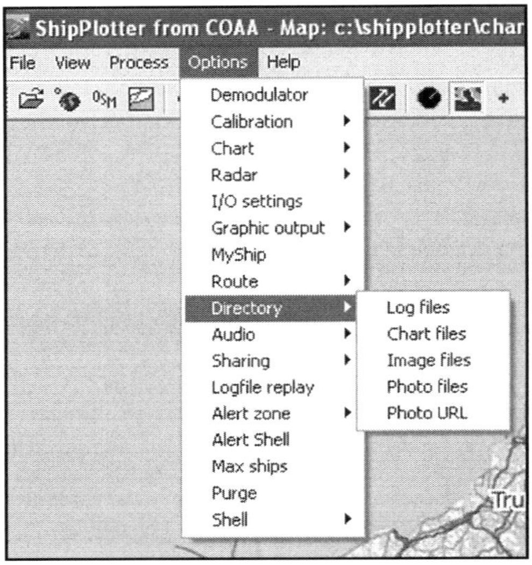

Figure 23 - Options, Directory

Figure 24 - Options, Directory, Chart Files

Chart files, Image files and Photo files checking that they all point to a folder off the main C:\ShipPlotter folder.

Selecting the first chart

Having created the first chart and defined where it is stored the user can now load the chart together with its calibration file into ShipPlotter by selecting *"File, Open new chart"* from the main menu bar as shown in Figure 25. Selecting the Chart_1 file will then load and display the file into the main ShipPlotter screen.

Figure 25 - File, Open new chart

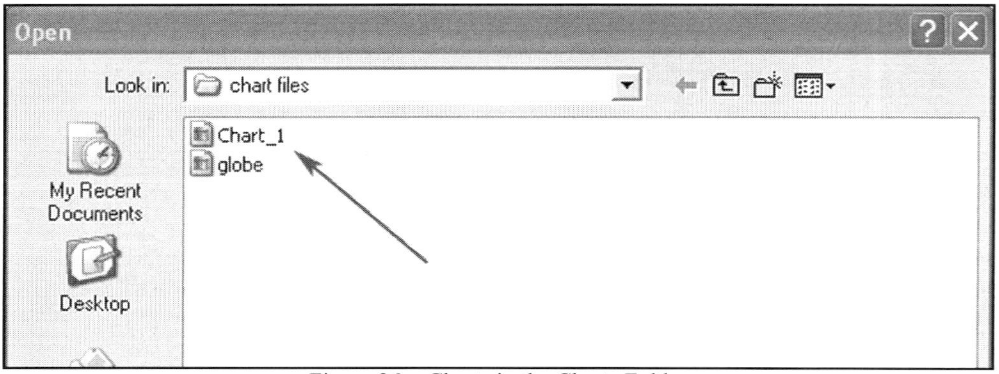

Figure 26 - Charts in the Charts Folder

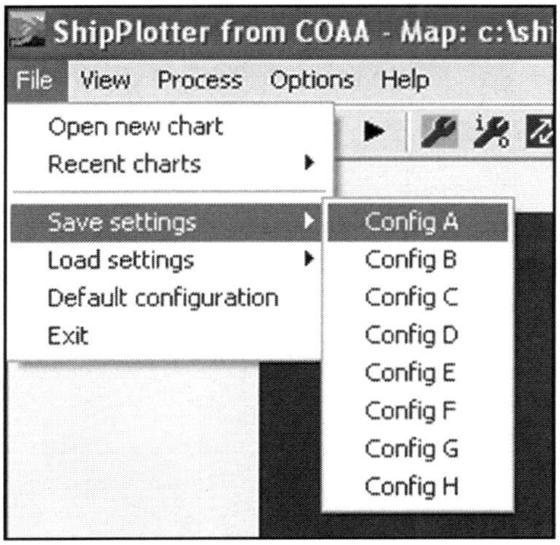

Figure 26 - Saving Config A

At this stage the user should save all the settings established so far by selecting 'File, Save Settings, Config A' as shown in Figure 26.

Now assuming that a radio as discussed in Chapter 3 has been installed, connected and is working the user can then click on the green "Start Button" to commence processing the received AIS data.

By clicking on the "Messages Button" the user can confirm that data is being received.

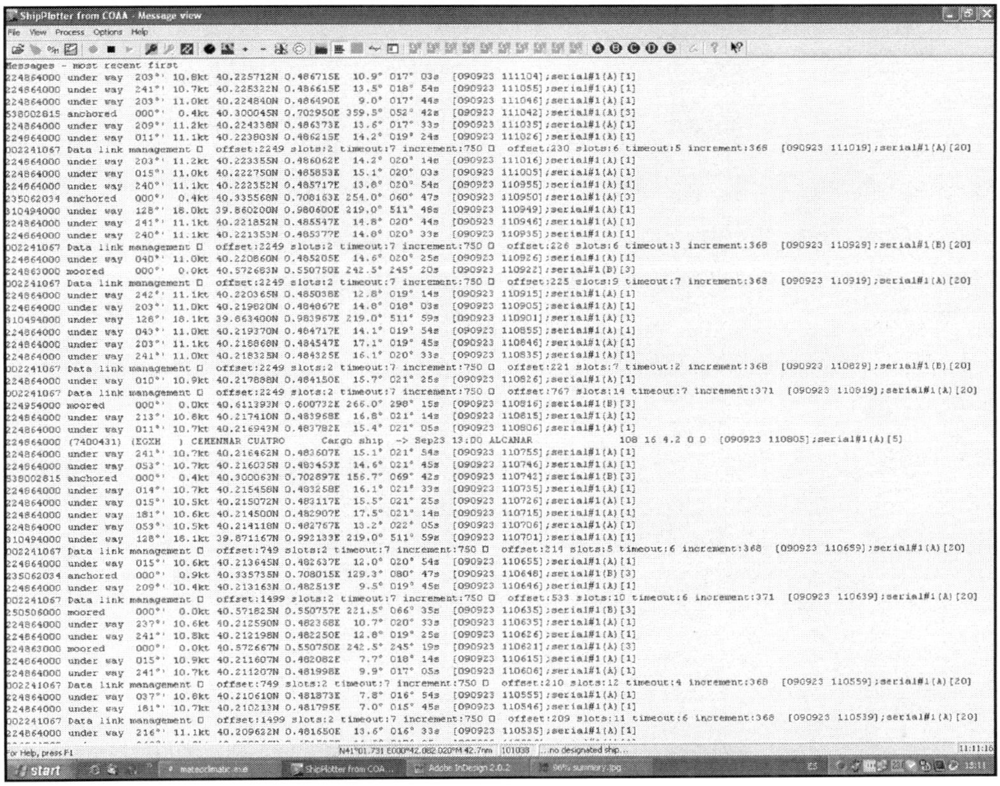

Figure 27 - Messages

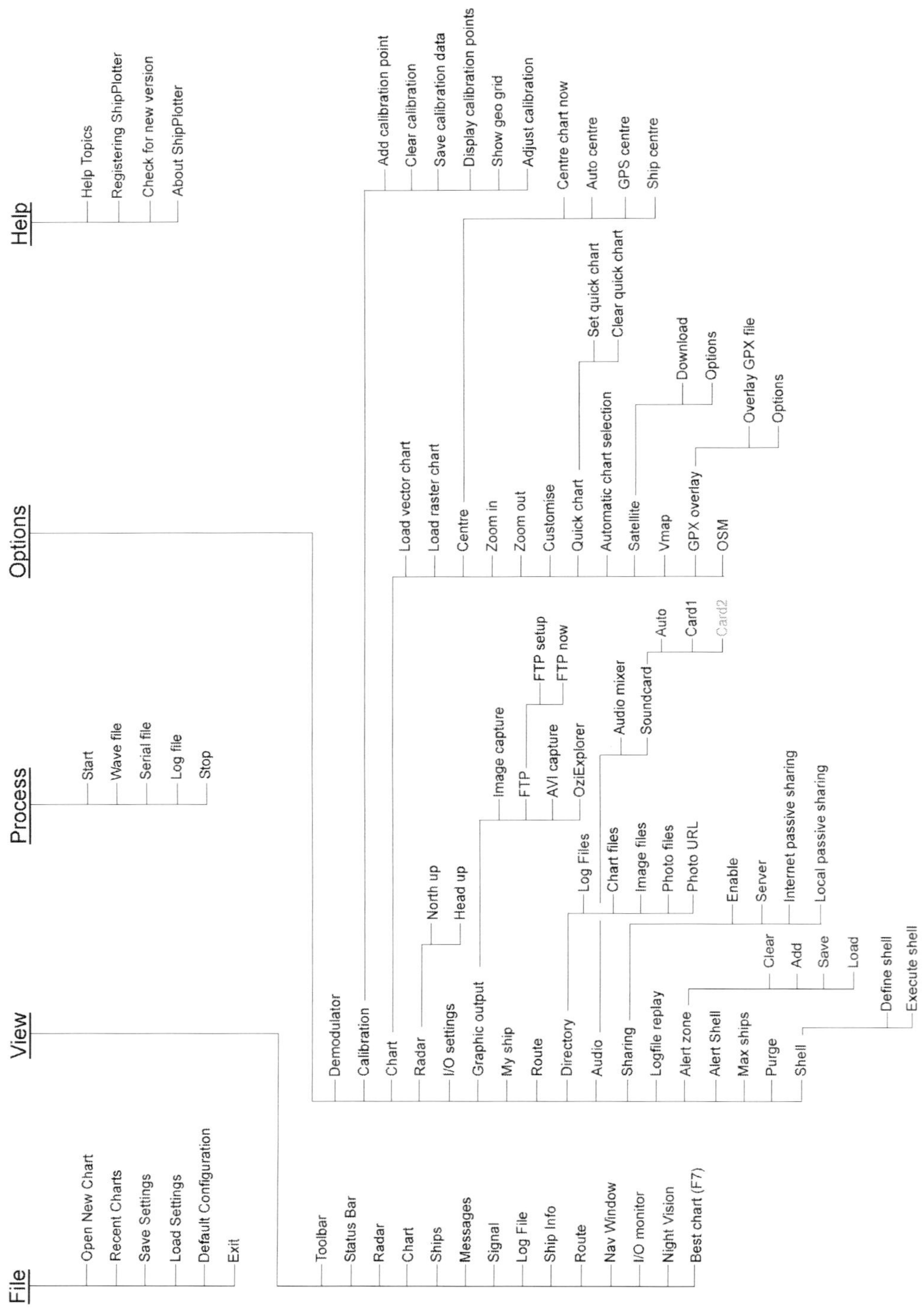

Figure 28 - ShipPlotter Menu Structure

Chapter 6

ShipPlotter Menus

As can be seen in Figure 28 ShipPlotter has a comprehensive Menu structure to deal with all the features and facilities available in the program. This chapter describes in detail each of the Menu commands.

File Menu Commands

The File menu offers the following commands:

Open New Chart

This command is used to load a new ShipPlotter chart background. Charts or maps are graphic BMP or JPG files and must be accompanied by a CLB calibration file.

Save Settings

This command is used to save the current configuration. All ShipPlotter settings and parameters (plotting colours, quick charts, current chart, chart position, chart scale, input/output options, registration state, alert settings, sharing settings, etc.) are saved by these commands.

The saved configuration includes the software registration number. If a configuration is saved before registering the software and then loaded again later it will lose the registration.

Load Settings

This command is used to load a previously saved configuration (plotting colours, quick charts, current chart, chart position, chart scale, input/output options, registration state, alert settings, sharing settings etc.).

Default Configuration

This command is used to reset all settings and configurations back to the original factory settings. A warning dialog will be displayed asking for confirmation that the user is prepared to lose all existing settings and data. To provide a limited fallback option, the current settings are all saved in the configuration H.

If the user wishes to revert to the settings before this option was selected they can do so by using the *"File, Load settings, Config H"* option.

N.B Deleting a configuration will also delete the "shipinfo file". It can only be recovered if it has been manually archived.

Exit

This command will end the current ShipPlotter session.

When the program exits, a file called *"shipinfo.txt"* is created with the static information for each ship plotted. When the program restarts, this file is reloaded so that ship names etc are restored from one session to the next. This process is completely automatic and transparent to the user. The file is also saved automatically at intervals when processing incoming messages. In this way, the recent data is preserved even in the event of an unexpected interruption of the program.

Dredger - Los Olivillos
Photo Courtesy and Copyright of Ken Anderson

View Menu Commands

The View menu offers the following commands:

Toolbar

This command is used to show or hide the toolbar at the top of the ShipPlotter screen.

Status Bar

This command is used to display and hide the Status Bar at the foot of the ShipPlotter screen. The Status Bar has five areas of interest. The left area of the status bar describes actions of menu items as the arrow keys are used to navigate through the menus or to describe the actions of toolbar buttons as they are pointed to. The second area shows the Geo Co-ords of the current cursor position. The next the number of messages received then the name of vessel bring tracked with its speed, course and time of last message and finally the current UTC time.

Radar

This command is used to show the position of ships on a radar-like display that is broadly similar to the Chart View except that no chart information is shown nor needed. The display is always centred on the origin of the range rings. This can be selected with the *"Options, Chart, Customise"* dialog.

Only the latest position from each ship is displayed. If the name of the vessel is known, derived from the appropriate message type, then the name is displayed. If the name of the ship has not been received only the MMSI number of the ship is shown together with the country of registration.

The user can zoom in and out of the display with the zoom buttons. At each zoom level, the program plots a series of concentric rings around the centre of the chart at intervals defined in the *"Options, Chart, Customise"* dialog. The odd 5NMs are plotted with a dashed black line and the 10NMs are plotted with a solid black line.

The information shown on the Radar display can be customised with the *"Options, Chart, Customise"* command.

For live data, ships that have reported within the last Grey timeout time are plotted in blue. Ships that have not reported for the Grey timeout time are plotted in grey. Ships that have not reported for the Omit timeout time will not be plotted although the data remains in the various log files and the *"View, Ship"* screen.

If a ship is moving, the recent positions are plotted behind the ship as a series of blue wake bubbles. Note that depending on the scale of the chart and the speed of the vessel, these bubbles may not be visible. The projected position in ten minutes time, based on the present course and speed, is indicated by a single line extension to the ship symbol. The tip of the line is the projected position of the ship.

Right clicking close to one of the ship symbols will pop up a dialog showing all the information available about the ship nearest the click point. This may include the MMSI number of the ship, the name, destination, ETA, position, course, speed and overall dimensions of the vessel depending on the types of message received as some of this information may not be available.

In the pop-up dialog there is a search button. If the computer is connected to the Internet then clicking on this button will initiate a search of the ITU web site for further details of the vessel based on the MMSI number.

Chart

This command is used to show the position of ships displayed on the currently selected chart.

If the name of the vessel is known, derived from the appropriate message type, then the name is displayed. If the name of the ship has not been received only the MMSI number of the ship is shown together with the country of registration.

The user can zoom in and out of the display with the zoom buttons. At each zoom level, the program plots a series of concentric rings around the centre of the chart at intervals defined in the *"Options, Chart, Customise"* dialog. The odd 5NMs are plotted with a dashed black line and the 10NMs are plotted with a solid black line.

The information shown on the Radar display can be customised with the *"Options, Chart, Customise"* command.

For live data, ships that have reported within the last Grey timeout time are plotted in blue. Ships that have not reported for the Grey timeout time are plotted in grey. Ships that have not reported for the Omit timeout time will not be plotted although the data remains in the various log files and the *"View, Ship"* screen.

If a ship is moving, the recent positions are plotted behind the ship as a series of blue wake bubbles. Note that depending on the scale of the chart and the speed of the vessel, these bubbles may not be visible. The projected position in ten minutes time, based on the present course and speed, is indicated by a single line extension to the ship symbol. The tip of the line is the projected position of the ship.

The user can move around the chart by left clicking on the chart and dragging it to the position that they want. This function is only available if auto-centring of the chart is not enabled. If auto-centring is enabled the chart is automatically positioned around the median of the received positions and clicking on the chart has no effect.

The information shown on the chart can be customised with the *"Options, Chart, Customise"* command.

Right clicking close to one of the ship symbols will pop up a dialog showing all the information available about the ship nearest the click point. This may include the MMSI number of the ship, the name, destination, ETA, position, course, speed and overall dimensions of the vessel depending on the types of message received as some of this information may not be available.

Ships

This command is used to display a line of data about every ship from which an AIS messages has been received.

This may include the MMSI number of the ship, the name, destination, ETA, position, course, speed and overall dimensions of the vessel depending on the types of message received as some of this information may not be available. It also includes the range from the current reference point which can be changed in the *"Options, Chart, Customise"* dialog. The last symbols on each line denote the source of the data. If the data was locally received, it will be '*' with an optional user-defined second symbol. If the data was received by sharing, the user who received the original data will be identified by a two-letter code.

The entry for each ship is updated with the latest position course and speed as more messages are received from the same ship.

Typically the messages that describe the position of the ship are sent more frequently than messages that give the full name, destination etc. of the ship. Until the full name of the ship is received, the MMSI number of the ship, in place of its name, identifies the position of the ship. ShipPlotter can be forced to use the full name of the ship even before that information is included in a received message by creating a file called "*shipnames.txt*" consisting of a series of MMSI numbers followed by a space, followed by the full name of the ship. This file should be stored in "C:\ShipPlotter\Log files" folder. When a message is received from a ship and this file exists it will be scanned to see if the MMSI is found. If so, the full name obtained from the file will be used on the chart. This view includes a vertical scroll bar so that the use can scroll through more than a screen full of ships.

Each line in the display is colour coded to show the status of the ship in the chart display.

Blue text denotes that the ship has a current report and is displayed.

Grey text denotes that the ship has not reported for the Grey timeout time and is displayed in grey on the chart.

Black text denotes that the ship has not reported for the Omit timeout time and is not displayed on the chart.

Green text denotes that the ship has been selected for unconditional display and will be displayed on the chart regardless of timeouts.

Red text denotes that the ship has been set to be never displayed and will not appear on the chart.

By clicking on a listed ship the display will change to the Chart View with the chosen ship centred in the chart. Right clicking on a listed ship will open Ship Information dialog box for that ship.

By clicking on one of the column headings in Ship View, the list of ships will be sorted in ascending order of that property. Right clicking on a column heading produce a sort in descending order of that property.

Messages

This command is used to display all the decoded AIS messages as they are received. The latest message is always at the top of the list.

Signal

This command is used to display the raw digital signal from the receiver in a graphical way. It is intended to be used only as part of the set-up process and is not useful once the system is working.

Log File

This command is used to display today's log file in a Notepad window.

Ship Info

This command is used to display the file "tempshipinfo.txt" in a Notepad window. The list may then be printed or saved as an archive.

Route

This command is not normally used for a shore-based user.

Nav Window

This command is not normally used for a shore-based user. However it opens a small window showing details of the designated ship and of its relationship to the active route point if any. Selecting the command a second time will close the window. Double clicking on it in the chart display can designate a ship. The data for the designated ship appears in the status bar of ShipPlotter. The designated status can be unset by double clicking a second time.

I/O Monitor

This command is used to display the Input Output Monitor window that provides a quick check on message activity on the various data inputs and outputs that ShipPlotter handles. Activity will only be seen on those inputs or outputs that are enabled in *"Options, I/O"* settings. Selecting the command a second time will close the window.

Night Vision

This command is used to change Windows system and background colours to suit operation of the program at night on board a boat or ship and would not normally be used for a shore-based user.

Best Chart (F7)

This command will automatically load the most suitable chart from those charts saved in the Charts folder for the currently received data.

High Speed Craft - Halunder Jet
Photo Courtesy and Copyright of Holger Jaschob

Process Menu Commands

The Process menu offers the following commands:

Start

This command is used to start the processing of input signals. Depending on the options made in the I/O settings dialog the ShipPlotter program will process either audio, serial data or take data from a remote source. The processing of audio data presents a heavy workload to the PC and it is not wise to run other processor intensive tasks at the same time.

The ShipPlotter program can start processing immediately by using command line with the argument '/start'.

Wave File

This command is used to start processing of recorded audio from a wave file. Note that the wave file must have been recorded as 8 bit mono data sampled at 48k samples per second.

Serial File

This command is used to start processing data captured from a serial port from a dedicated AIS receiver to a chosen text file. The serial data contained within the selected text file will be processed until the file is exhausted. The chart, ship and message views can be used during and after processing. This process proceeds very much faster than real capture time. A progress strip allows the user to replay the file at various speeds and to pause the replay or abandon it.

Log File

This command is used to start processing data that has been captured previously and saved in a log file. ShipPlotter does this automatically.

Data contained within the selected log file will continue to be processed until reaching the end of file marker. Chart and ship views can be used during and after processing.

This process proceeds very much faster than real capture time thus providing a high-speed replay of data previously received and logged. It is a useful feature if wanting to review activity if ShipPlotter for example has been left running overnight.

Replay of a log file can be performed at various speeds, or paused, using the buttons in the progress window. The replay starts in the Paused state thus one of the following options must be selected to begin the replay.

The **"Step"** button advances the log file by a small amount and then pauses.

The **"RT"** button replays that log file at the original rate at which the log file was captured. The program notes the time tag of the first entry in the log file and then allows the replay to proceed such that the replay elapsed time matches the progression of real time.

The **"Slow"** button advances the log file continuously at a slow rate.

The **"Fast"** button advances the log file continuously at a higher rate.

The **"Adv.To"** button lets the user enter a specific time (in UTC). The replay will quickly advance to that time and then pause.

Stop

This command stops any processing currently in progress.

Passenger Ship - The World
Photo Courtesy and Copyright of Ken Anderson

Options Menu Commands

The Options menu offers the following commands:

Demodulator

A dialog box is displayed to set the options for how the data demodulator operates.

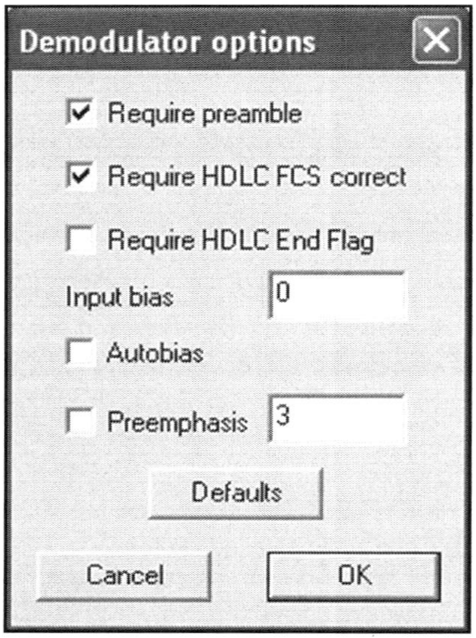

Figure 29 - Deomodulator Options

Require preamble

AIS messages must have a preamble in accordance with the AIS specification. The default condition is for this option to be enabled.

Require HDLC FCS correct

The messages transmitted by AIS equipped ships include a checksum. If this option is checked, only messages with a valid checksum will be displayed. The default condition is for this option to be checked.

Require HDLC End Flag

The last part of the digital message is an end flag. The default condition is for this option to be unchecked.

Input bias

If the display in the Signal view is persistently offset from the centre line of the chart, putting a small non zero value into the Rx Bias may improve system performance. The default value is zero.

Autobias

If this option is checked, the mean signal level is determined from the preamble and is used in the demodulation of the rest of the message. This option may help with some receivers and soundcards, especially if there is some doubt about the frequency accuracy of the receiver. The default value is disabled.

Preemphasis checkbox

Enable this option to add a FIR filter preemphasis to the incoming audio processing. This option adds additional processing load but, depending on the users receiver, it may considerably increase the message decode success rate. The next field defines the slope of the preemphasis.

Preemphasis parameter

This parameter determines the slope of the preemphasis applied to the incoming audio. A value of zero means that the FIR filter is flat.

Experiment with values in the range 2-10 to see which one gives the highest number of decoded messages. Values outside of these limits are unlikely to improve the performance of the decoder. The optimum value will depend on the receiver being used. The default value is 3.0.

Default button

The default button restores the default values to these options.

Calibration

To use a chart or map as the base for the graphic display in ShipPlotter, it must first be calibrated so that the software correctly associates chart pixels with geographical coordinates.

Add calibration

This option is used to add a new calibration point. When the red target toolbar button is depressed, the next mouse click over the area of the chart is taken to be a definition of a calibration point.

Clear calibration

This command is used to delete all calibration points for the current chart. Once the calibration points are cleared the chart may be calibrated again.

Save calibration data

This option is used to save the calibration data generated for the current chart. If the data is not saved then the work done in calibrating the chart for use with ShipPlotter will be lost. The calibration file must be saved with the same prefix as the associated map file but with the extension ".clb".

Display calibration points

This option is used to display red target symbols on the current chart showing where the calibration points are.

Show geo grid

This option will overlay a geographical grid on the current chart. This option is useful for verifying the quality of the calibration. Ideally the red crosses will lie directly over the corresponding fiducial marks on the chart. The number of fiducial marks is chosen to suit the scale of the map. They are plotted at 1 minute, 10 minutes, 1 degree or 10 degrees intervals according to the map area displayed.

Adjust calibration

This option is used to correct small displacements in the calibration data for the current chart. Such errors may result from a difference in the map datum used to calibrate the chart or may arise from the slight differences in projection or scale that come with downloaded charts.

Chart

There are a number of different options for ShipPlotter charts.

Load vector chart

This command loads a new raster chart from a graphics file. The required file is selected in the file dialog that appears. The view mode will change to Chart mode when a new chart is loaded.

Load raster chart

This command loads a new vector chart from a VEC graphic file. The required file is selected in the file dialog that appears. The view mode will change to Chart mode when a new chart is loaded.

Centre

There are four chart-centring options.

Centre chart now

With this option selected in Chart mode, the current chart is automatically centred

Auto centre

This option centres the chart display to the median position of the current ship positions.

GPS centre

With this option selected the chart will be centred to the current position reported by a local GPS receiver. This option is only enabled if enabled GPS input is enabled in the “Options, I/O” settings dialog.

Ship centre

With this option selected the chart is centred on the current position reported by the designated ship (the last ship that was double left clicked on). As the ship reports new positions, the chart position will change so that the ship remains in the centre of the screen.

Zoom in

This command is used to zoom into the chart. The displayed scale is increased by 1.4 times for each click so that the scale doubles for every two clicks. The mouse wheel can also be used to zoom in and out.

Zoom out

This command is used to zoom out of the chart. The displayed scale is decreased by 1.4 times for each click so that the scale halves for every two clicks. If sharing is enabled then the shared data will automatically be refreshed to collect data for the newly expanded area, which may take a few seconds. When zooming out with the mouse wheel the automatic sharing refresh can be disabled by holding down the Ctrl key at the same time as using the mouse wheel.

Customise

This command is used to configure the display of data on the chart. See Chapter 8 Charts & Calibration for more detailed information on customising charts.

Quick chart

There are two options for Quick charts.

Set quick chart

This option enables each of ten quick chart buttons on the toolbar to be assigned a specific chart.

Clear quick chart

This command is used to delete a quick chart.

Automatic chart selection

This option is used to make ShipPlotter automatically change charts if a ship that is being tracked moves off the current chart. A ship must first be designated by double clicking on the ship; a ship-centred display must have been selected. If the ship moves off the current chart ShipPlotter will then search the calibration data of all the charts in the chart directory to find the one most suitable.

Satellite

There are two options for Satellite.

Download

Before this option is used a chart must already be displayed on the screen. This option is used to download a satellite image of the area currently displayed on the screen. The new image will be automatically calibrated and it will be placed in the current Chart Directory with a file name based on the current date and time.

Options

This option displays a dialog that allows the type of image that is downloaded to be selected. The image type options they are only applicable to the first two satellite sources. The first four are

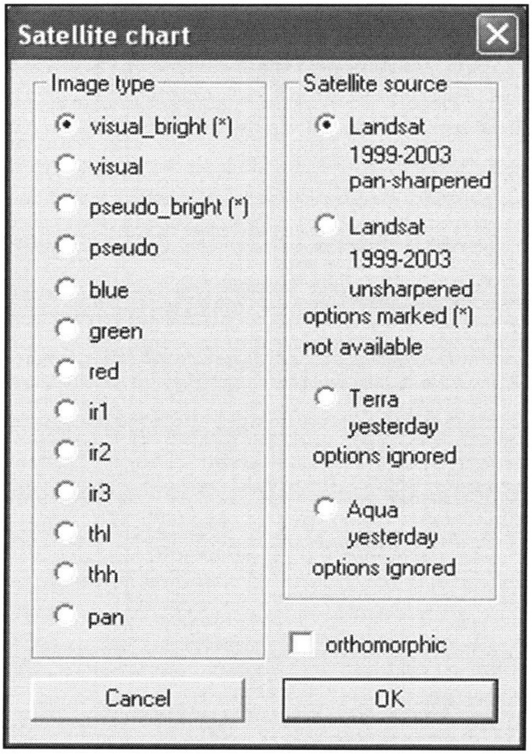

Figure 30 - Satellite Chart Options

coloured and the remainder are monochrome. The "visual_bright" and the "pseudo_bright" options are not available in the unsharpened dataset.

The Landsat images were captured between 1999 and 2003. The default pan-sharpened image provides the best resolution.

The Terra and Aqua satellites provide lower resolution images but are more topical as they are images captured in the last day or so. As they are recent images they will reflect the cloud and ground cover at the time they were captured.

The Orthomorphic option forces the downloaded satellite image to have the same horizontal and vertical scales locally. Enabling this option avoids the situation where a large area chart imposes that distortion on satellite images that are generated from a zoom in to such a chart. Because of the change in relative scales, the resulting satellite image chart will not cover precisely the same geographical area as the parent chart.

Vmap

This command is used to create an outline chart of the area currently displayed on the screen using data from VMAP level zero. The data is contained in charts with the extension ".vec" which are compressed versions of selected layers from the NIMA VMAP data. The default chart included with ShipPlotter shows the coastline of Western Europe. Other "vec" charts for selected areas may be downloaded from the COAA web site at http://www.coaa.co.uk. The feature is provided to give a limited coastline and location data in areas where better charts may not be to hand.

GPX overlay

There are two GPX overlay options.

Overlay GPX file

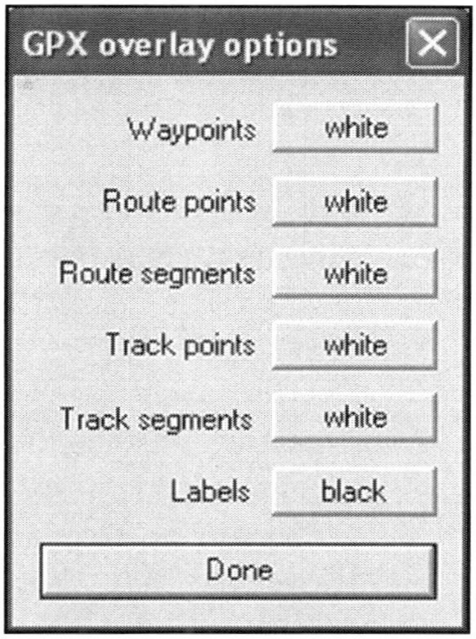

Figure 31 - GPX Overlay

ShipPlotter can superimpose waypoints, tracks or routes on top of any chart. A file of such data must first be converted to GPX format. This is an open XML format generated directly by many GPS management programs. If a GPS uses a proprietary format for tracks, routes and waypoints they can probably be converted to GPX format using one of the freeware GPS utilities such as GPSBabel or EasyGPS.

Track points are plotted with a small triangular symbol, which is filled with the user-selected colour and surrounded with an outline in the currently defined text colour. A straight line joins successive track points in a user-defined colour. None of the nodes are labelled.

Waypoints are plotted with a small triangular symbol, which is filled with the user-selected colour and surrounded with an outline in the currently defined text colour. Each waypoint is labelled if a name is included in the GPX file. ShipPlotter uses the same font size, colour, background, case rules and opacity for labelling waypoints as are defined for ship labelling.

Route points are plotted with a small triangular symbol, which is filled with the user-selected colour and surrounded with an outline in the currently defined text colour. A straight line joins successive track points in a user-defined colour. Each route point is labelled if a name is included in the GPX file. ShipPlotter uses the same font size, colour, background, case rules and opacity for labelling route points as are defined for ship labelling.

Note that displaying a route with this option is not the same as using a route for navigation. Routes displayed with this function are purely cosmetic.

Options

This command is used to select the colours for the symbols and the joining lines.

OSM

This option will download a map of the area covered by the current display from the OpenStreetMap server. This is perhaps the best chart option for the shore based ShipPlotter.

Radar

This command is used to show the position of ships on a radar-like display. The display is always centred on the origin of the range rings as selected with the *"Options, Chart, Customise"* dialog. This can be a fixed position such as the Home Station location or a designated ship.

North up

This command orientates the Radar display as a standard compass North Up.

Head up

This command orientates the Radar display with top of the screen the direction that a designated ship is heading.

I/O settings

This command is used to load a new ShipPlotter chart background. Charts or maps are graphic BMP or HPG files and must be accompanied by a CLB calibration file.

Graphic output

There are four Graphic Output options.

Image capture

This option can be used to save a screen capture, typically of the chart display, to a designated named file at regular intervals. The file will be saved in the "C:\ShipPlotter\Image files" folder. Not the image size will be the same as the screen size as it will not be scaled.

A filename and how regularly the image will be saved are entered in the dialog box. The Data time annotation option adds a date timestamp in UTC or Local Time to be overlaid on the image.

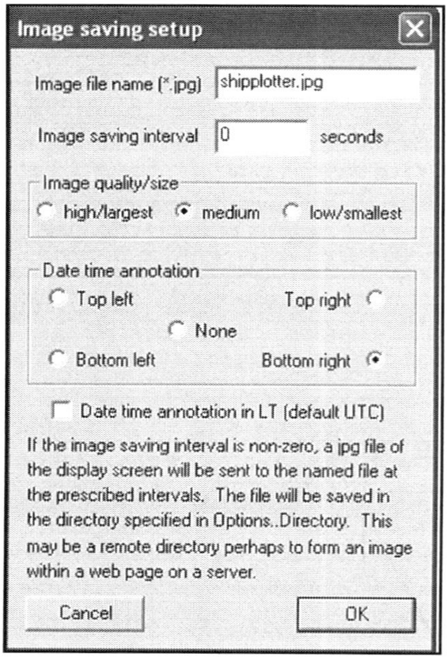

Figure 32 - Image Capture Dialog

FTP

The FTP options allow the user to upload an image files by FTP to a web server.

FTP setup

Selecting FTP Setup displays a dialog box where the URL of the server, a username and password can be entered. The Directory entry must be a valid accessible folder on the host server.

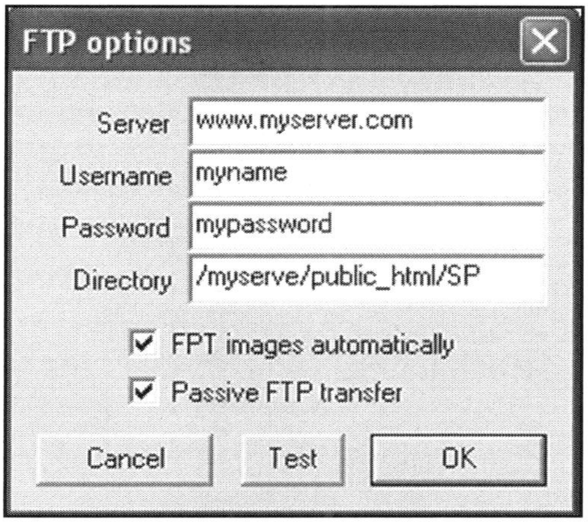

Figure 33 - FTP Setup

If the 'FTP images automatically' option is checked then an FTP transfer will take place any time an image is generated by the *"Options, Image capture"* as described on the previous page. Even with the 'FTP images automatically' option checked the FTP will not take place unless automatic image saving is also enabled in the "Options, Images setup" dialog and current date is being processed.

The Test button can be used to start an immediate FTP transfer and test that the settings are correct. Progress of the transfer will be displayed in the title bar of the application.

FTP now

This command is used to start an immediate FTP transfer and test that the settings are correct. Progress of the transfer will be displayed in the title bar of the application.

AVI capture

With this option selected ShipPlotter will save a succession of screen images to frames to an AVI file thus creating a time-lapse sequence in movie format

Both the recording rate, specified as the number of real-time seconds between successive frames, and the playback rate, the number of frames per second when the AVI movie is replayed, can be defined.

It is recommended not to make the recording interval too short as the recording of each frame takes a significant amount of time and the real-time processing of signals could be severely curtailed.

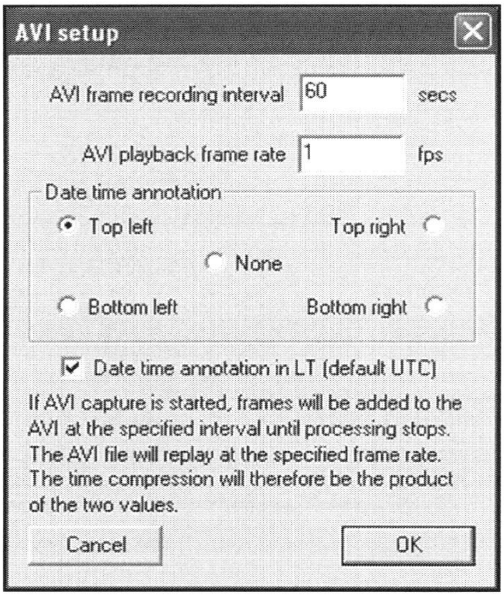

Figure 34 - FTP Setup

At most map scales there will not be much movement visible in under a minute so a recording interval around that time scale is a good starting point. Any playback rate can be used but for large image sizes the playback data rate may be too high for the processor. To start off with saving frames every 60 seconds and a playback rate of 1 fps is recommended as the time-lapse speed up will still be 60 times, a ten knot ship will going at 600 knots.

The Data time annotation option adds a date timestamp in UTC or Local Time to be overlaid on the image. Separate menu commands allow the user to start and stop the AVI recording while ShipPlotter is processing data.

OziExplorer

This command is used to start transmission of ship position data to the OziExplorer mapping program provided it is installed and running. When processing live data the ship positions will be sent to OziExplorer once per minute and plotted on the current OziExplorer map as " OziEplorer Map Features".

My ship

This option is not required for a shore based ShipPlotter. The option is for a small boat owner that is not equipped with an AIS transponder. It is used to define a name and MMSI number to be attached to 'My Ship' that is plotted at the position derived from the small boats GPS.

The name defaults to "My Ship" and the MMSI number to 999999999. The Ship name must be less than 20 characters and must consist of letters only. The MMSI number must lie between 999999000 and 999999999 to avoid confusion with real MMSI numbers.

The Ship Type can be selected from 100 predefined types. The option "Plot my ship" determines whether or not this MMSI number will be plotted. By default, the option is checked so that "My Ship" appears on the chart. For most applications, this option should be checked. The Destination can be any string of up to 20 characters and the ETA can be any string of up to 11 characters.

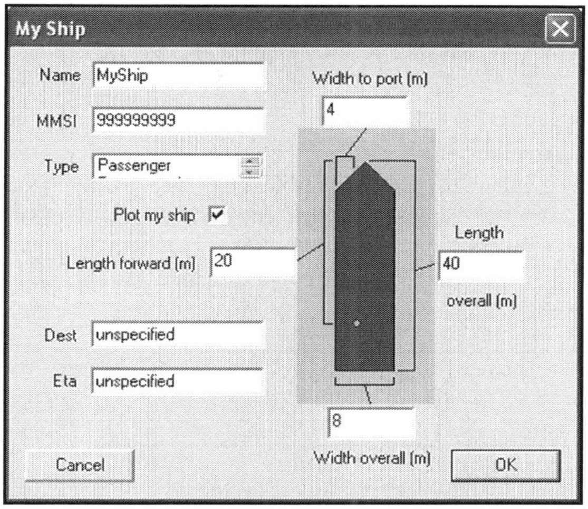

Figure 35 - My Ship

Route

Again this option is not required for a shore based ShipPlotter. A route is a sequence of geographical points. ShipPlotter allows the user to create, edit, save, load and use routes for tracking a vessel's progress.

Directory

See the earlier basic options section in Chapter 5.

Audio

If a standard radio received or scanner is being used to receive the AIS signals the Audio options enable the user to set up the parameters of an audio mixer and/or sound car.

Audio mixer

The content of this dialog depends on a particular sound card. The inputs used should be the "Line In" or "Aux" ones.

Soundcard

The PC may be fitted with a number of different types of soundcard.

Auto

This command automatically determines the best soundcard for ShipPlotter to use.

Card1

Checking this option forces the use of a particular soundcard.

Sharing

Sharing enables users all over the world to share received data with one another.

Sharing is discussed in greater detail later in Chapter 9.

Enable

This command enables the transfer of shared data.

Server

This command defines the parameters of the web servers used to transfer date.

Internet passive sharing

This options allows the user to passively receive data from the Internet without having to process it

Local passive sharing

This option allows the user to passively receive data without having to process it but from another copy of ShipPlotter running on the same computer or on another computer running on a local area network.

Logfile replay

This command is used to replay the data previously saved in a Log file.

Alert zone

An Alert Zone can be defined to provide a warning to the user of any ship entering the zone.

Clear

This command is used to clear or remove a previously placed alert zone from the chart.

Add

This command is used to create a new Alert Zone. With this option selected clicking on the chart in a systematic manner vertex points will be added creating a polygon defining the Alert Zone. Up to 99 points can be added.

Once the Alert Zone polygon has been created the *"Options, Alert Zone, Add"* option should be clicked again to reset the option otherwise additional clicks will add more vertices to the polygon

Save

This command is used to save the currently displayed zone in the Log Files folder. A number of different Alert Zones can be created and saved under different filenames all ending with the suffix ".lrt"

Load

This command is used to load a previously saved Alert Zone file from the Log Files folder.

Alert shell

This command opens a dialog four different commands can be defined that will then be executed by the system when an alert occurs.
The categories are:

Alert on proximity to the reference position.

This option is defined and enabled in the *"Options, Chart, Customise"* dialog.

Alert on a new MMSI (not previously received).

This option is enabled in the *"Options, Chart, Customise"* dialog.

Alert on a listed MMSI.

The listed MMSIs are defined and the option is enabled in the *"Options, Chart, Customise"* dialog.

Alert on a position report from within an Alert Zone.

This alert occurs when a condition is met within the Alert Zone option is enabled in the *"Options, Chart, Customise"* dialog.

Alert on a route event.

There are two route events that cause an alert. See "Options, Route, Alert" for more details.

When one of these alerts occurs, provided it is enabled, the corresponding command will be executed. ShipPlotter creates a text file called alertmessage.txt that contains a text description of the nature of the alert including the MMSI number of the ship involved.

The default shell commands are a set of batch files invoking the "Blat.Exe" utility that automatically sends an email to a specified email address. Each of the predefined batch files makes use of the "alertmessage.bat" message that was described above. Further information on "Blat.exe" can be found at http://www.blat.net. As an alternative to using "Blat.exe" the user can create and use any other batch file to carry out any required function that will be performed when an alert condition occurs.

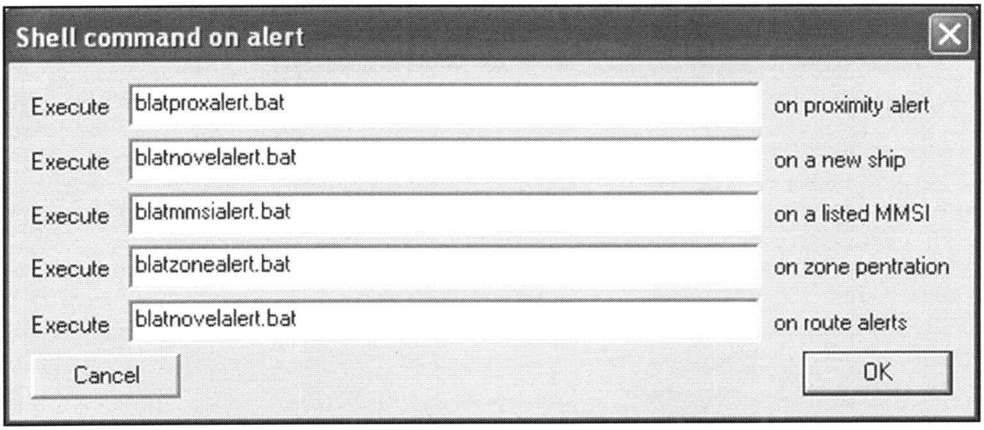

Figure 36 - Alert Shell Dialog

Max ships

This option is to specify the maximum number of ships in the internal database. The default value is 25000. If the database is resized the current ships are purged but the static information is preserved.

Purge

This option can be used to clear all messages from the database. It also removes all ships from the Chart View and from the Ship View but leaves all the static information such as names etc still in the database.

Shell

There are two options here.

Define shell

This option is used to define the Shell function that can be executed with either the Execute Shell command or by clicking the ShipPlotter Toolbar Shell button.

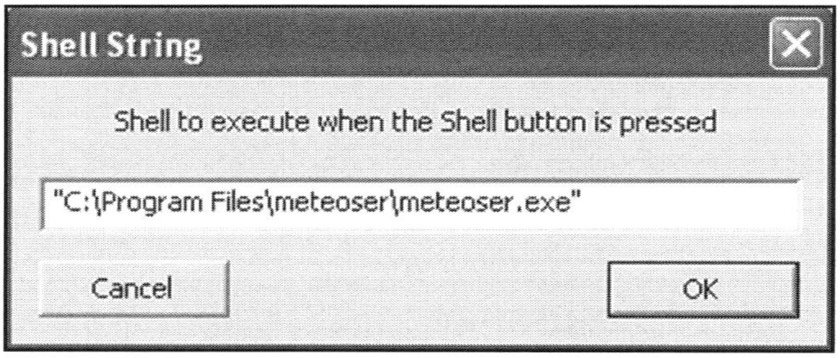

Figure 37 - Example Shell String

A Shell function is a software program, a command, a script or a website URL that can be executed by the computer at any time without closing the current ShipPlotter session.

Execute shell

This command will immediately execute the Shell command.

Help Menu Commands

The Help menu offers the following commands:

Help Topics

This command is used to launch ShipPlotters' online help and tutorial system.

Registering ShipPlotter

This command is used to register the current installed copy of ShipPlotter.

Check for new version

This command is used to check for a newer version of the software.

About ShipPlotter

This command is used to display the copyright notice and version number of the installed copy of ShipPlotter.

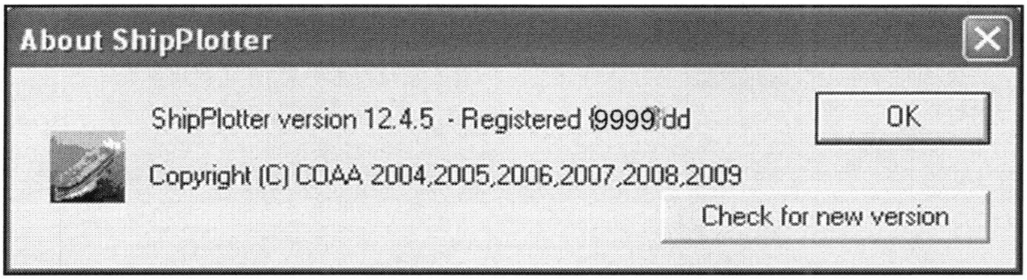

Figure 38 - About Box

Steam Ships (Operating and Preserved) - Boroysund
Photo Courtesy and Copyright of K Mikkelsen

Chapter 7

The ShipPlotter Toolbar

The toolbar is displayed across the top of the ShipPlotter application window and below the menu bar. The toolbar provides quick mouse access to many of the tools used in ShipPlotter and complimenting the ShipPlotter menus.

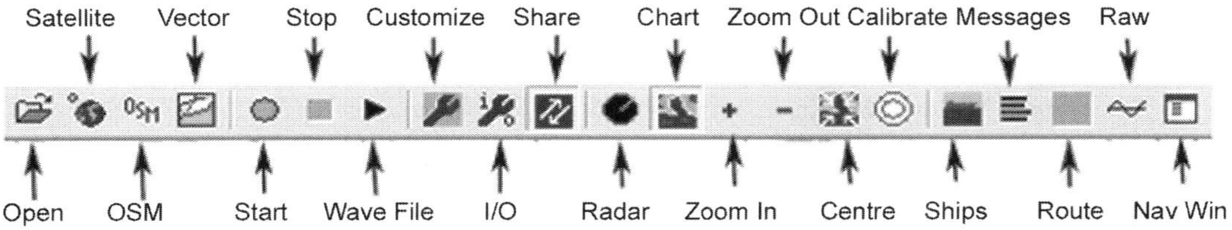

Figure 39 - The ShipPlotter Toolbar

File
Opens a chart file

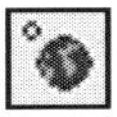

Satellite
Downloads a satellite chart, to replace the current chart, of the type defined in *"Options, Chart, Satellite, Options"* as described on Page 43 in Chapter 6.

OSM
Downloads an Open Street Map, to replace the current chart.

Vector
Replaces the current chart with a vector outline map.

Start
Starts the processing of ships data.

Stop
Stops processing data.

Wave File
Starts processing a pre recorded wave file.

Customize

Clicking this button opens the Plotted Information window that is used to set up the parameters of what is displayed on a chart.

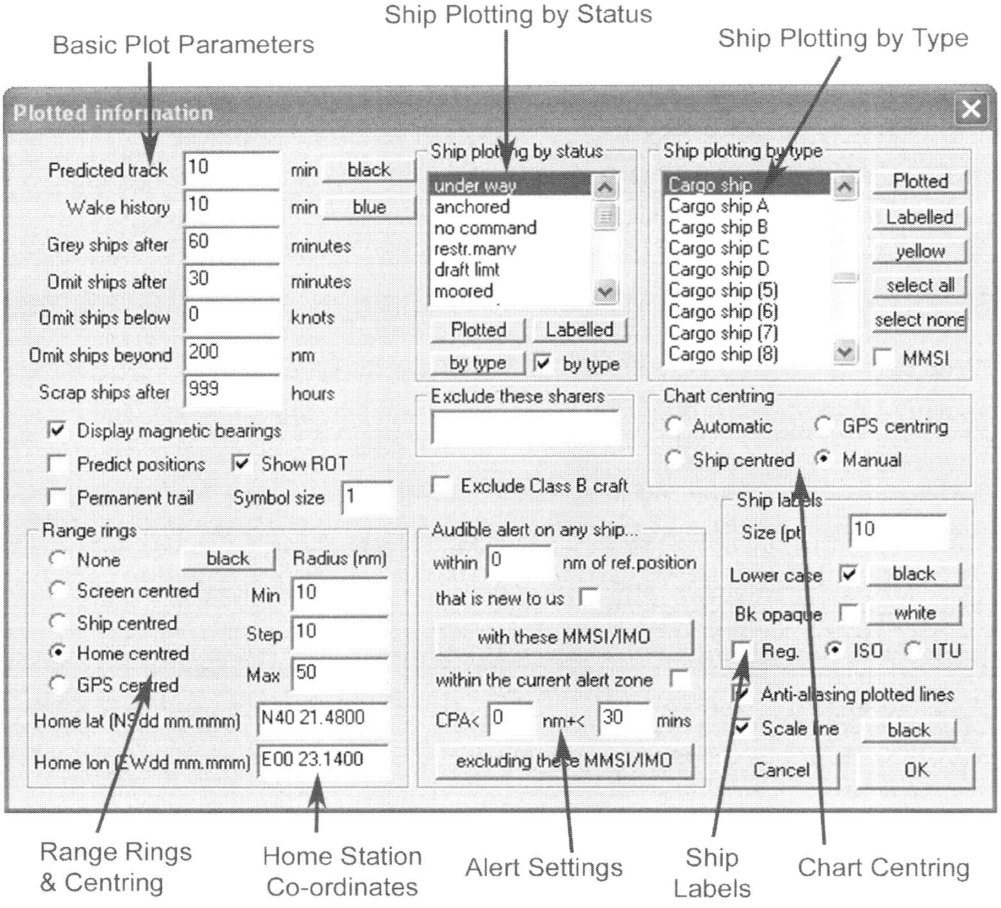

Figure 40 - Customize Plotted Information Dialog

Basic Plot Parameters

This group is used to set up the basic tracking parameters.

Predicted track

This value determines whether and by how much, the current position of the ship is extrapolated. The prediction line is based on the reported course-over-the-ground, which may differ from the reported heading. If the ship reports its rate of change of heading, then the track is curved to show the way the ship is turning.

Setting the value to zero will not display the prediction lines. The button alongside the track enables the colour to be used for the vector on the chart to be selected.

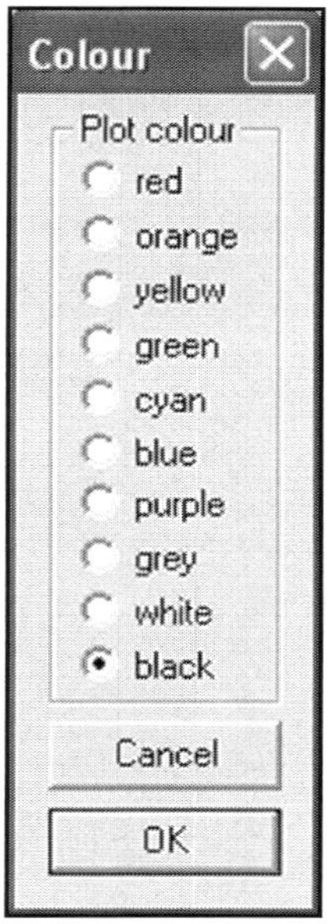

Figure 41
Colour Selector

Wake history

This value determines how many previous position reports are plotted as a blue wake behind the ship. The maximum is twelve minutes and ten reports. The length of the wake on the chart depends on the scale of the chart, the speed of the vessel and the frequency of position reports. If the chart covers a large area, the ship is reporting frequently and moving slowly, not all of the position reports will be included in the wake history. The button to the right of the Wake history option allows the user to customise the colour used to draw the wake. The default colour is blue.

Grey ships after

This value sets the time after which a ship report is greyed out. If a report is not received from the ship for more than this time, the ship symbol and its associated wake, will be plotted in pale grey.

Omit ships after

This value sets the time after which a ship report is not plotted at all. If a report is not received from the ship for more than this time, the ship remains in the Ship View but it is not plotted on the chart.

Omit ships below

This value allows the exclusion of slow moving vessels from the chart. This option is used with a very small value to exclude vessels that are moored or anchored but whose status is still "Under way".

To show all ships, regardless of speed this value should be set to zero.

Omit ships beyond

This value allows the omission of ships beyond a certain range from the defined Home Station location.

This can be used to avoid unreliable intermittent plots of very distant ships.

Setting this number to a large value will force all ships to be included.

Scrap ships after

This value enables ShipPlotter to discard all data about a ship including the static information such as ship's name etc after the specified number of hours has elapsed without a report being received. The value 999 hours is a dummy value meaning "Never discard the ship's data".

Display Magnetic Bearings

ShipPlotter normally displays all courses, headings and bearings in degrees True (°T). Checking this option will display bearings in degrees Magnetic (°M).

Predict positions

Checking this option will extrapolate ship positions based on the speed and course that the ship is reporting and the time since the last position report. The true reported position is drawn in outline and the extrapolated position is shown in the appropriate colour for the ship type, etc.

Show ROT

If this option is checked, the predict vector is curved to show the rate of turn (ROT) of the vessel. Note that not all vessels provide this information.

Leave a permanent trail

With this option checked ship positions are recorded by a black dot that remains on the chart until such time as the chart is reloaded. This is independent of the dynamic wake plotting options. The trail only shows on parts of the display that correspond to the source chart, it does not record on any parts of the display that are off the edges of the current chart.

Range rings

These options control the plotting of concentric range rings on the chart. The user can select to have no range rings or to have them centred.

Screen centred

The range rings are centred on the middle of the screen.

Ship centred

Clicking on a ship is clicked displays the range rings are centred on that ship.

Home centred

The range rings are centred on the Home Station location.

The user can choose at what intervals the rings are plotted and what minimum and maximum radius should be plotted. The colour of the range rings by pressing the colour button in this group of parameters.

Home Station Co-ordinates

The Home Station location should be entered in degrees, minutes, and decimals of a minute of latitude and longitude with the quadrant in capitals preceding the degrees, for example N40 21.4800 and E00 23.1400.

Ship Plotting by Status

The Ship Plotting by Status group enables the user to include or exclude ships on the chart on the basis of the reported current status. The default is to display and label all statuses and to colour the ship symbols according to type. When one of the statuses in the list box is selected the plotted, labelled, and colour buttons will show the current plotting parameters for that status. Clicking on the "plotted" or "labelled" buttons enables the condition relating to the currently highlighted status to be changed.

Ship Plotting by Type

This group enables the user to include or exclude ships on the chart based on their type as reported in the static message. The default is to display all types. Note that static messages are transmitted less often than position messages so the ship type will default to 'unspecified' until a static message is received. Checking the MMSI checkbox causes ShipPlotter to label ships by their MMSI number instead of their name. Again clicking the colour button can change the colours.

Exclude sharers

If using Internet sharing of data, the user can specifically exclude messages from certain other sharers by putting their code letters in this box. The sharer identification code letters are displayed in the pop-up flags that appear when the cursor is placed over a ship in the chart display. For reasons of upward compatibility, there are two formats that can be used.

Exclude sharers – no commas

An exclude list without commas "," is interpreted as a list of the initial share code letters of the sharers. For example, the exclude sequence "abcde" will exclude all ships contributed by any sharers whose sharing identity begins with “a”, “b”, “c”, “d” or “e” regardless of the second letter. This would, for example, exclude both sharer ‘a0’ and ‘aA’.

Exclude sharers – with commas

An exclude list that includes at least one comma is interpreted as a list of sharing codes including both symbols if there are two letters/symbols between the commas or the first sharing letter if there is only one symbol between the commas. Thus the exclusion list “ab,cd,e” will exclude sharers with the sharing code “ab”, “cd” and any sharer whose first letter is “e”. To exclude just one sharer using both parts of the sharing code, a comma must be appended to the exclusion list in order for the two symbols to be interpreted as the complete sharing identity and not as two separate single letter exclusions. Thus an exclusion list “ab” will exclude all sharers whose identity begins with “a” or “b” while the exclusion list “ab,” with a comma will only exclude the sharer with the identity “ab”.

Ship Labels

The text label for a ship can be customised to some extent with this option. Checking the “Reg” box enables/disables the display of the country of registration with the ship name.

The choice of ISO or ITU determines what abbreviations on the chart label are used for the country of registration when the ship name is not available. The ISO option is a two-letter code that is also used for Internet domains for example United Kingdom = “UK”. The ITU option is, in most cases, a three-letter code for example Canada = “CAN” but there are some exceptions. For example the United Kingdom is simply “G” using the ITU code.

Audible Alerts

This group of options defines the various alert settings.

Alert on any ship within ‘x’ of the reference position

An audible warning will sound if a ship is within the entered range of the reference point. A red warning will appear at the top of the window and will stay there until the warning is acknowledged by clicking on it.

Alert on any ship that is unknown to us

An audible warning will sound whenever a message is received from a vessel that has not previously reported.

Alert on the ship with MMSI/IMO number

An audible warning will sound if a message is received from any ship with a specified MMSI or IMO number.

Alert on any ship within the current alert zone

An audible warning will sound and a warning message be displayed whenever a ship reports a position inside the alert zone polygon.

Alert on any ship with a CPA

Audible warnings will sound for ships entering the Closest Point of Approach (CPA). A red warning will appear at the top of the window and will stay there until the warning is acknowledged by clicking on it.

Alert excluding these MMSI/IMO numbers

Audible warnings will not be generated for ships with a specified MMSI or IMO number.

Anti-aliasing plotted lines

With this option checked, the predict line, wake lines and ship outlines are all plotted using an anti-aliasing algorithm which makes them look less jagged.

Scale line

The distance scale is a short vertical line that is superimposed on the right hand edge of any chart view or radar view to indicate the map scale at the current zoom setting.

Chart centring

This group of options determines how the range rings are centred on the screen.

Automatic

The chart is automatically centred on the centre of gravity of all ship positions.

Ship centred

Clicking on a ship centres the chart on that ship.

GPS centred

The chart is centred on the boats GPS position.

Manual

Dragging the cursor on the screen centres the chart.

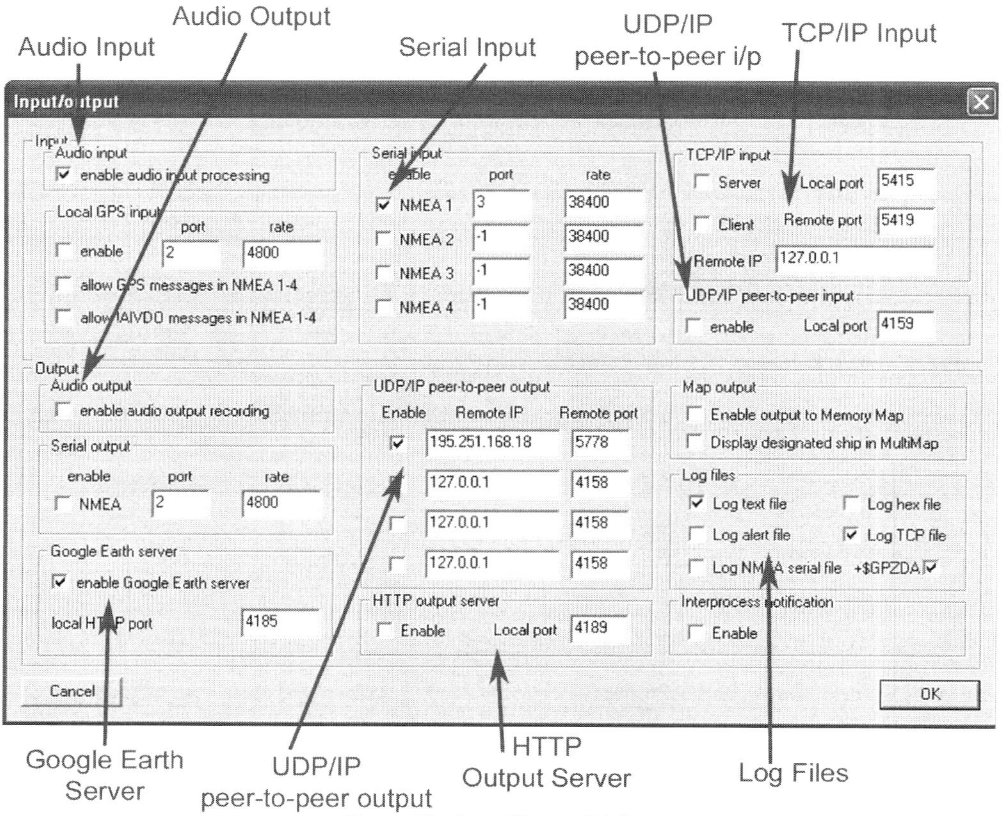

Figure 42 - Input/Output Dialog

I/O

Clicking this button opens the Input/Output window that is used to set up the parameters of the source of ships data input to ShipPlotter and processed data output from it. This button is greyed out when ShipPlotter is processing data. To reactivate the button the "Stop" button must first be pressed.

Audio Input

This box should be checked when processing signals from a VHF receiver. The enable option has three states:

i. Box blank - no input

ii. Box checked - input with the possibility of sharing the input data

iii. Box checked in grey - input with sharing disabled

Audio Output
When processing audio input a wave file of the received audio can be saved by checking this option. This option is not available if the Audio Input is not checked.

Serial Input
ShipPlotter has the ability to process serial data from up to four dedicated AIS receivers that each provides a serial output. To use a serial input enable option should be checked and a Com Port number and the data rate specified. Options not used should be unchecked. Each of the four NMEA input options has three states:

i. Box blank - no input
ii. Box checked - input with the possibility of sharing the input data
iii. Box checked in grey - input with sharing disabled

In addition to sharing, the condition also controls whether or not serial data is sent to the serial data output when enabled or to the NMEA format log file when enabled.

UDP/IP peer-to-peer Input
ShipPlotter can receive data from ShipPlotter running on another computer using UDP/IP over the Internet or a LAN. The UDP/IP input enable option has three states:

i. Box blank - no input
ii. Box checked - input with the possibility of sharing the input data
iii. Box checked in grey - input with sharing disabled

In addition to sharing, the condition also controls whether or not UDP/IP data is sent to the serial data output when if enabled or to the NMEA format log file when enabled.

UDP/IP peer-to-peer Output
ShipPlotter can send received messages to up to four other computers each running the program using UDP/IP.

TCP/IP peer-to-peer Input
Data from a remote TCP/IP server or client such as a serial to TCP/IP adaptor connected to a serial AIS receiver at a remote site. Both TCP/IP input options have three states:

i. Box blank - no input
ii. Box checked - input with the possibility of sharing the input data
iii. Box checked in grey - input with sharing disabled

In addition to sharing, the condition also controls whether or not TCP/IP data is sent to the serial data output when if enabled or to the NMEA format log file when enabled.

HTTP Output Server

ShipPlotter can act as an HTTP server to provide ship data to other computers running the program.

Google Earth Server

Checking this option enables Google Earth to displays the ships currently processed by ShipPlotter.

Log Files

This group has several options. Log files are saved in the "C:\ShipPlotter\Log files" directory.

Log text file

The user can save the received message data as a text file that can be replayed at a later date.

Log hex file

The data can be saved as a hex file. This option can be used for diagnostic and development purposes.

Log tcp file

TCP/IP messages can be saved in a log file.

Log alert file

Alert messages can also be saved in a log file.

Log NMEA Serial File

Decoded messages can be saved as a log file in serial NMEA format. This log file can also be replayed as it retains all the message data from the incoming data but does not contain any timing information so the replay will not be time tagged.

Share
Clicking this button starts the data sharing processes previously set up in the *"Options, Sharing"* menu item.

Radar
Clicking this button changes the display from a chart to show the position of ships on a radar-like display. The display is always centred on the origin of the range rings as selected with the *"Options, Chart, Customise"* dialog. This can be a fixed position such as the Home Station location or a designated ship.

Chart
Clicking this button changes the display to the default chart previously set up.

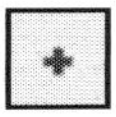

Zoom In
Clicking this button zooms in on the displayed chart or radar type display.

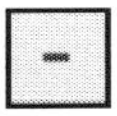

Zoom Out
Clicking this button zooms out of the displayed chart or radar type display.

Centre
Clicking this button centres the chart as set up in *"Options, Chart, Centre"* dialog.

Calibrate
Clicking this button starts the chart calibration process as described more fully in Chapter 8 - Charts & Calibration.

Ships
Clicking this button changes the display to provide a line of data about every ship whose AIS messages have been received. This may include the MMSI number of the ship, the name, destination, ETA, position, course, speed and overall dimensions of the vessel depending on the types of message received as some of this information may not be available. Each line in the display is colour coded to show the status of the ship in the chart display.

Blue text denotes that the ship has a current report and is displayed.

Grey text denotes that the ship has not reported for the Grey timeout time and is displayed in grey on the chart.

Black text denotes that the ship has not reported for the Omit timeout time and is not displayed on the chart.

Green text denotes that the ship has been selected for unconditional display and will be displayed on the chart regardless of timeouts.

Red text denotes that the ship has been set to be never displayed and will not appear on the chart.

By clicking on a listed ship the display will change to the Chart View with the chosen ship centred in the chart. Right clicking on a listed ship will open Ship Information dialog box for that ship.

By clicking on one of the column headings in Ship View, the list of ships will be sorted in ascending order of that property. Right clicking on a column heading produce a sort in descending order of that property.

Messages

Clicking this button will display a list of all the decoded AIS messages as they are received. The latest message is always at the top of the list.

Route

This button may be greyed out, as it is not normally used for a shore-based user.

Raw

Clicking this button will open a graphical display to display an analogue representation of digital signal from the receiver. It is normally used as part of the set-up process and is not useful once the system is working.

Nav Win

This is not normally used for a shore-based user but clicking this button will opens a small window showing details of the designated ship and of its relationship to the active route point if any. Selecting the command a second time will close the window. Double clicking on it in the chart display can designate a ship. The data for the designated ship appears in the status bar of ShipPlotter. The designated status can be unset by double clicking a second time.

There are a second group of toolbar buttons to the right of the first set.

Quick Charts

There are a total of eight Quick Chart buttons. To start off with these buttons are normally greyed out. When a chart is allocated to one of the buttons the button is coloured. When the mouse pointer is hovering

over an allocated Quick Chart button the name of the saved chart will be displayed.
Note that saved Quick Charts will be saved with the current configuration. If a new configuration is opened it will not load the previously saved Quick Charts.

Config
There are a total of five Configuration buttons labelled A, B, C, D & E. Once a configuration has been saved it may be quickly recalled by clicking one of these buttons.

Shell
Clicking this button will initiate a previously defined shell command. If no shell command has been defined the button will be greyed out.

Help
Clicking this button will launch the Online Help and Tutorial screen.

About
Clicking this button will display the copyright notice and version number of the installed copy of ShipPlotter.

Tanker (Gas) - Gas Legacy
Photo Courtesy and Copyright of Aleksi Lindström

Chapter 8
Charts & Calibration

It is important to remind the reader that this book had been written for the shore based ship-spotting enthusiast and not for the mariner or small boat owner. Thus this chapter on Charts and Calibration is aimed at providing the user with charts suitable for use on their computers that have reasonable detail and accuracy for the type of user who remains firmly on shore in their armchair.

Charts

To use a chart or map as the base for the graphic display in ShipPlotter it must first be in the correct format and then it must also be calibrated so that the software correctly associates chart pixels with geographical coordinates.

ShipPlotter uses charts in the familiar JPEG graphics file format as well as commercial BSB charts. ShipPlotter can also download satellite images from the Internet. BSB charts and satellite images are self-calibrating. However other charts must first be calibrated as discussed later.

If a paper chart is used as the source ShipPlotter chart it can be scanned in and then edited with a standard graphics editor program. The graphics program should be used to size and orientate the chart to fit the users screen and be legible. Once this has been done the file should be saved in the "C:\ShipPlotter\Charts" folder as a JPG file.

An alternative to scanning a paper chart is to use a screen capture program to capture a chart image from a mapping program such as Google Maps, Multimap or better still Google Earth. A word of caution however – charts prepared from Google map and Multimap images may be difficult to calibrate, as they will have been stretched in one direction or the other to make them more suitable for road navigation and they ignore the curvature of the earth.

A better alternative is to download a prepared chart from a web site such as http://www.selkieman.co.uk/coggie/charts or by using a chart generator such as that found at http://emit.demon.co.uk/map2.php.

Calibrating the chart

To calibrate a graphical chart file it must first be loaded into ShipPlotter. Then a number of reference points must be chosen and marked on the chart. Finally the calibration data that has been created must be saved in a calibration file with the same name suffix as the original chart.

Load the chart

Select *"File, Load new chart"* and choose one of the JPG files that have previously been created. If the chart has not already been calibrated then a warning message will be displayed to that effect as this means that the program did not find a calibration file with the same name in the Charts folder.

If there is already a calibration file for this folder the user can proceed straight to the "Check the calibration" step.

Explore the chart

The first time that the chart is displayed it is scaled so that the whole of the chart is visible in the screen window. The user can zoom in and out of the chart using the Zoom buttons. [+] [-]

The user can also move around the chart by clicking on points anywhere in the window.

Calibrate the chart

The user should first use the zoom and centre buttons and controls to find a pace on the chart where the geographical coordinates are known precisely. A useful program for finding the geographical co-ordinates of places "Lat-Long Finder" can be found at http://www.satsig.net/maps/lat-long-finder.htm. Now click on the Calibrate button and then click on the chosen place. A dialog box will be displayed where the X and Y pixel coordinates on where the user clicked with two edit boxes for the latitude and longitude of the place. The user should then enter the two values for the place using the format of a letter for N, S, E or W and then two groups of digits. Note these values must be *degrees, minutes and decimals of a minute* and **not** *degrees, minutes and seconds.*

This process should be repeated for several points, a minimum of four, around the chart with them well spaced around the chart.

Check the calibration

Once the calibration points have been added the user should select "Calibration, Display, calibration points" to display the added calibration marks. If there are any errors these can be deleted by clicking on the calibration point in question, then checking the "delete" box and then clicking "OK".

Save the calibration.

When satisfied with the calibration process the user should select *"Calibration, Save calibration data"* to save the calibration file in the C:\ShipPlotter\Chart files" folder. The default file name will be same as the chart but with the file extension ".clb". e.g. If the chart name is "mychart.jpg" the calibration file must be "mychart.clb".

N.B. If the calibration data is not saved then all of work done in calibrating the chart will be lost when the program is shut down.

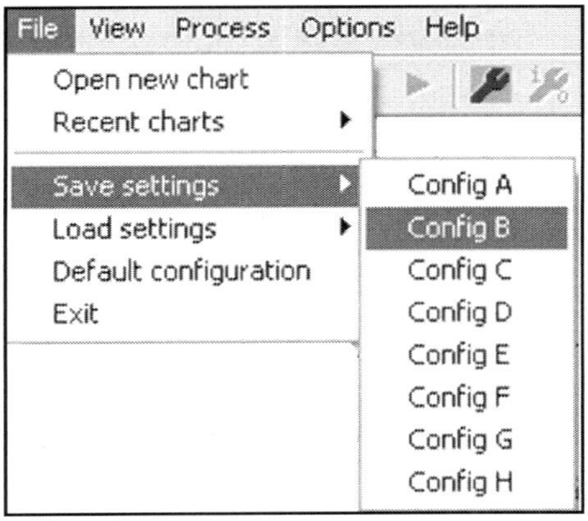

Figure 43 - Saving a Configuration

As well as saving the calibration file in the Charts folder it is important to save the current configuration when any changes are made using the *"File, Save, Config x"* menu option as shown in Figure 43.

Once a configuration has been saved it can be quickly recalled by clicking the appropriate "Config" button.

Note that if any charts have been saved as "Quick Charts" they will be saved with a Configuration and then recalled when that Configuration is reloaded.

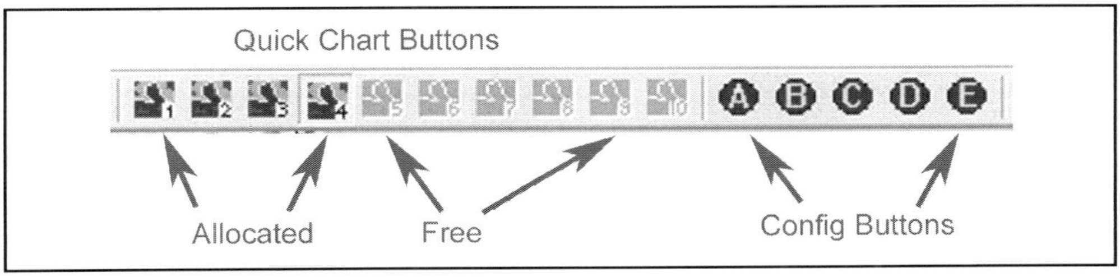

Figure 44 - Quick Chart Buttons

Example Charts

The next three pages show examples of charts all based on having created the first chart using the chart generator at http://emit.demon.co.uk/map2.php. This is shown in Figure 45.

The chart is for a shore based ship-spotting enthusiast living on the Mediterranean coast near Valencia in Eastern Spain. This is a reasonably busy shipping area with a mix of tankers, cargo ships and passenger traffic moving along the Spanish coast as well as to and from the Balearic Isles.

Once the initial chart and its calibration file have been saved in the Charts folder the chart is saved as “Config A”. At the same time the chart is also saved as “Quick Chart 1”.

By clicking the “OSM” (Open Street Map) 

button the chart shown in Figure 46 can be produced covering the same geographical area. The chart is then saved as “Quick Chart 2”.

Next clicking the “Satellite” button  produces yet another chart as shown in Figure 47 and again covering the same geographical area. This is then saved as “Quick Chart 3”.

After again saving as “Config A” the user can view the ships in the shipping area with three different chart views.

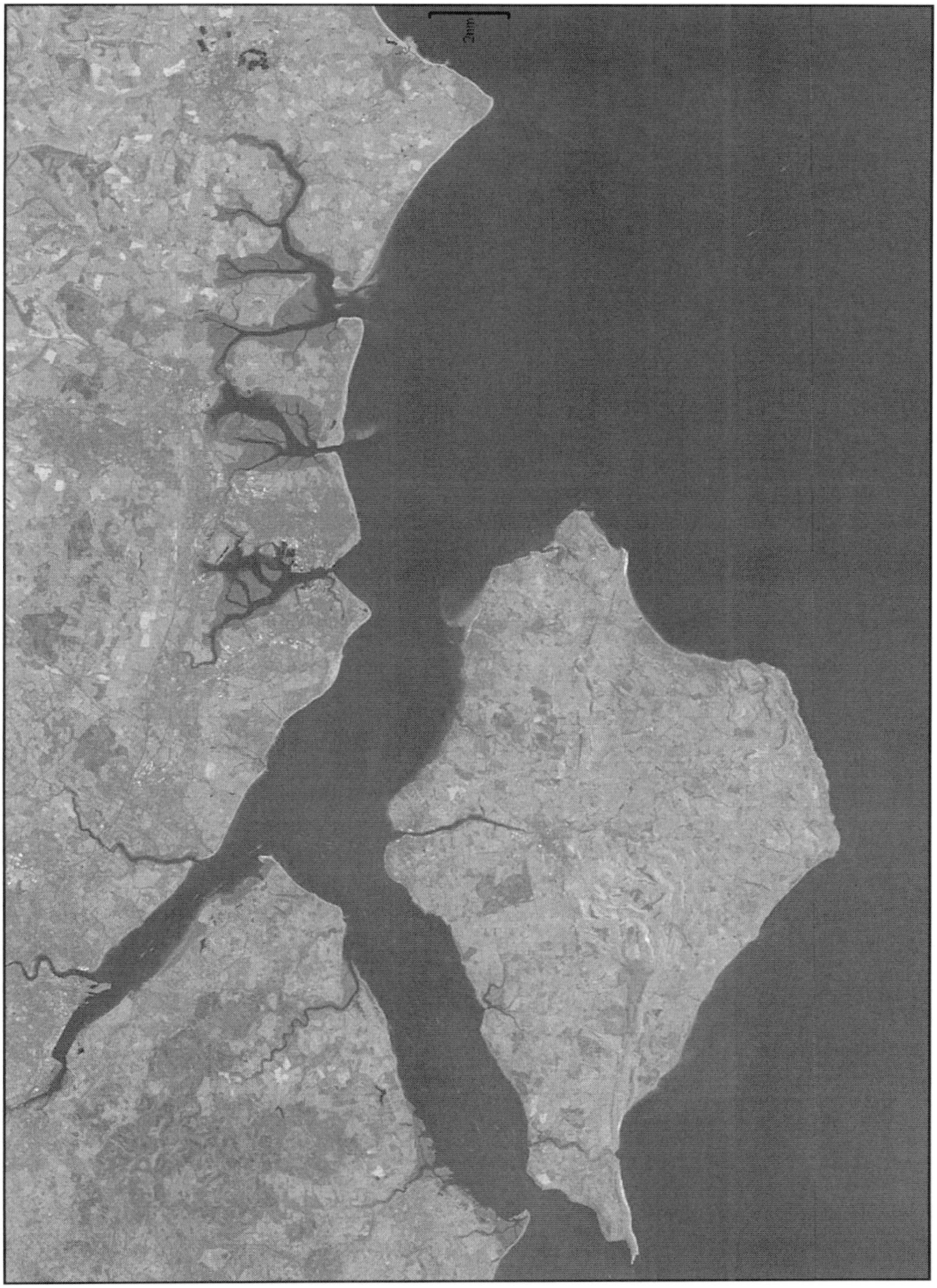

Figure 45 - Chart generated from http://emit.demon.co.uk/map2.php

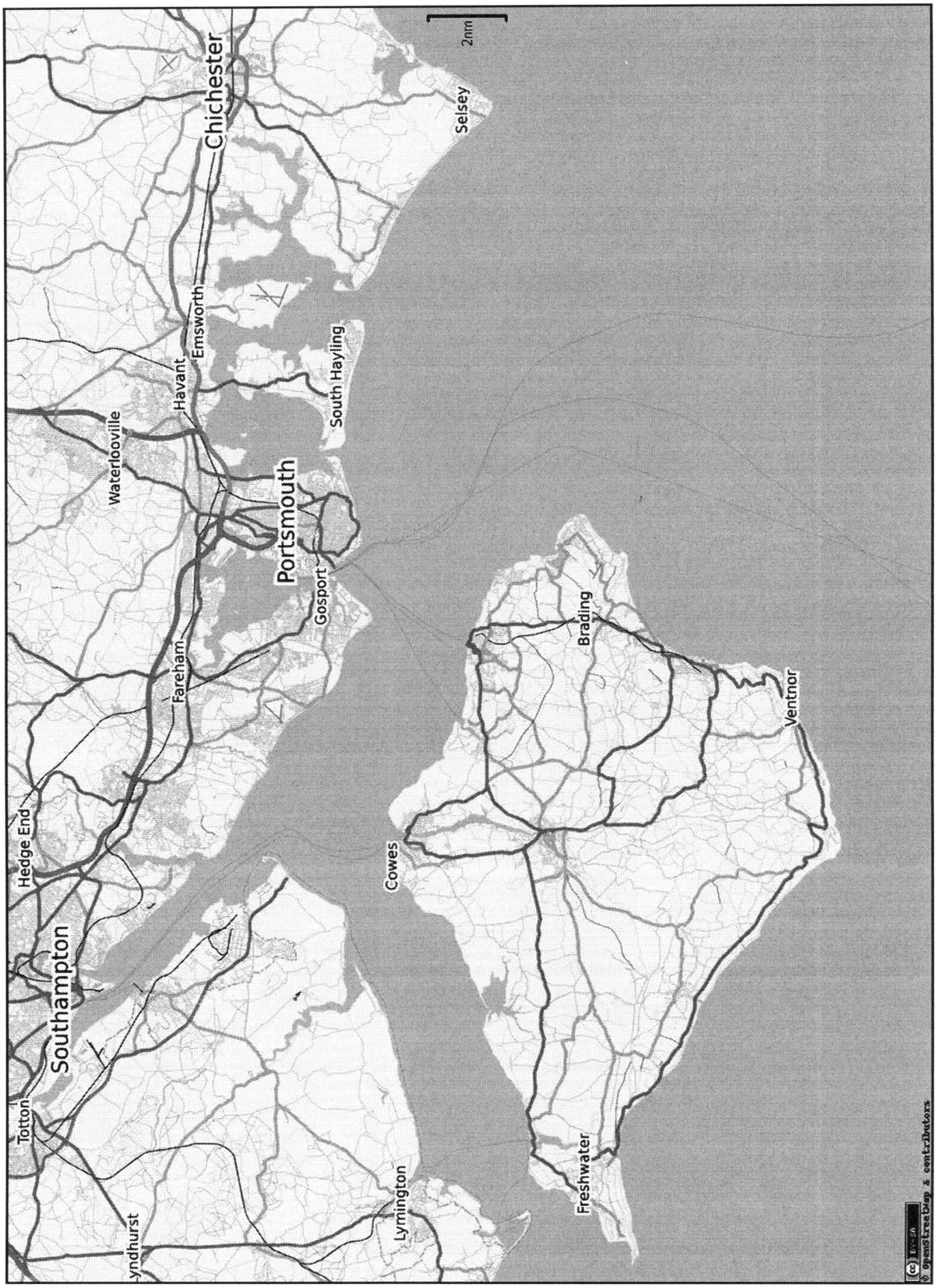

Figure 46 - OSM version of chart generated from http://emit.demon.co.uk/map2.php

Figure 47 - Vector map of chart generated from http://emit.demon.co.uk/map2.php

Tug - Arild Viking
Photo Courtesy and Copyright of Capt Jan Melchers

Chapter 9
Sharing

ShipPlotter users can share data in a different number of ways with other users using the Internet to transfer the data. For the shore based user the common method is for individual users to upload their own received data to a web server and to download data already uploaded by other users. This allows users to view shipping movements in many different parts of the world and not just their own area.

Groups of users say along a waterway or coastline can share their data to provide an overview of their area whilst not needing access to other parts of the world.

Sharing Via A Server

There are several ShipPlotter servers around the world readily available to users wanting to share data. Access to these servers is over the Internet and

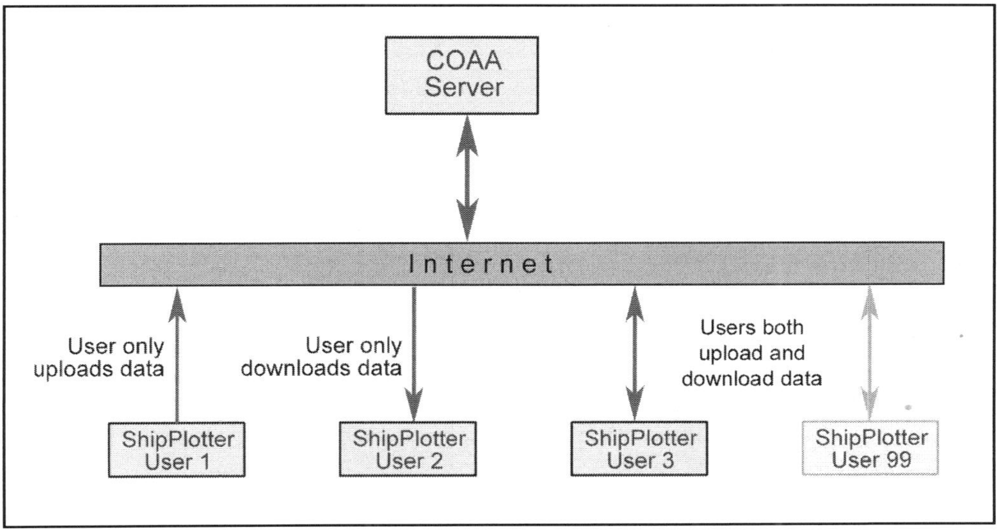

Figure 48 - COAA Server

they rely on the data uploaded being regularly updated by the subscribing users.

The COAA Server

The most widely used server is the "COAA Server" and this is currently available without charge to all ShipPlotter registered users. To set up sharing using the COAA Server first select from the Menu bar *"Options, Sharing, Server"* to open the dialog box as shown in Figure 45.

Here the user can specify two different remote servers with either or both be enabled or disabled

To use the COAA server the user should first select *"Options, Sharing, Server"*. Then check the "Enable sharing" box. If the user wishes to only send their own data to the COAA and not download the already shared data then the "Upload only" box should be checked. The server address should be set to www.coaa.co.uk and the script "shipinfo.php". The port number should be set

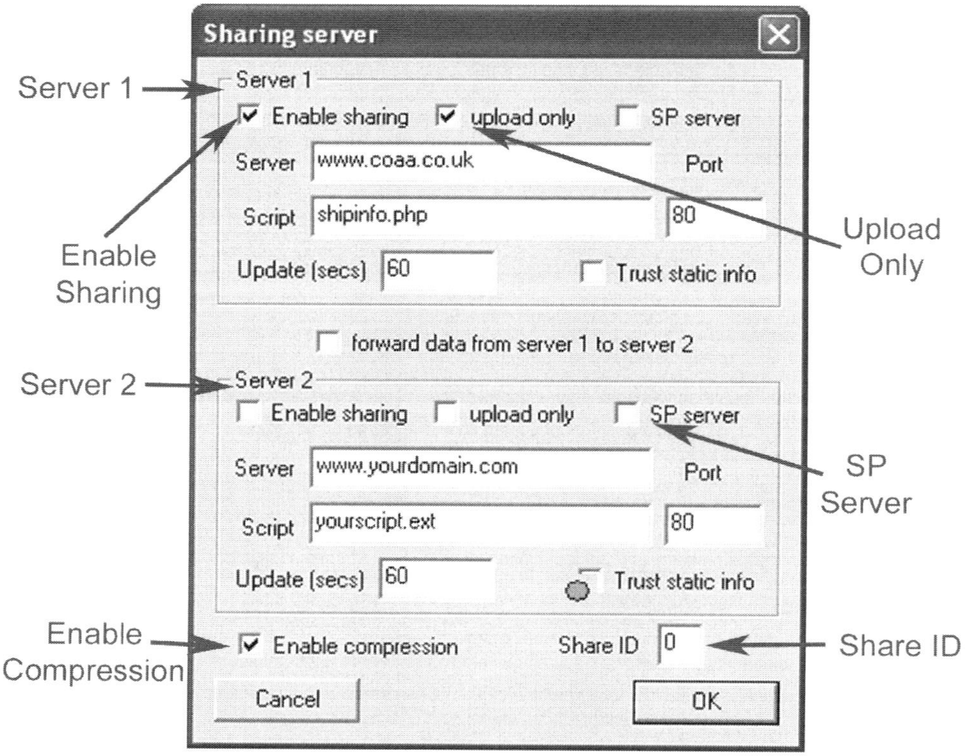

Figure 49 - Sharing Server

to 80. Next set the update period to 60 seconds and uncheck the "Trust static info" box.

Finally ensure that "Enable compression" box is checked as this allows the COAA server to compress the ship data that is downloaded to the program during sharing, and then click on the "OK button" to save the settings.

Start Sharing

To start sharing the user should ensure that the data is being processed by having first clicked on the Run button which should then be greyed out and then selecting "Options, Sharing, Enable" or clicking the "Share button". 

Note: If downloading data from the COAA Server then the "Enable audio input processing" box in the "Input/Output" must be checked, see Figure 40.

Once sharing is enabled the user should see in the lower left corner of the ShipPlotter screen messages *"Sharing upload 1"* followed by *"Sharing download 1"* followed by *"Sharing download 1 completed"*. Any ship reports that have been submitted to the COAA server by other users that are in the area of the users chart currently displayed will then also appear on the chart.

For those users not interested in seeing data from other users but wanting to make theirs available to others the "Upload only" box in the dialog should be checked.

Troubleshooting

If sharing cannot be enabled then either the copy of ShipPlotter is not registered or the settings in the *"Options, Demodulator"* dialog are incorrect, as ShipPlotter will not allow users to share data if the checksum checking in that dialog is disabled.

If no other users' ships are visible when sharing is enabled and processing is running there are a number of possibilities. These are that any other user sharing their data does not cover the chart area currently viewed. The chart is not correctly calibrated. The computers clock or time zone is incorrectly set so that the server rejects submissions as being outside of the permitted time window. Or that the computers firewall is blocking ShipPlotter from accessing the Internet.

Peer to Peer Sharing

Another method of sharing data is to send UDP/IP messages from one copy of ShipPlotter to another avoiding the need for a common server. Peer-to-peer connections can only transmit data in one direction although ShipPlotter can be configured to send data to up to four other machines.

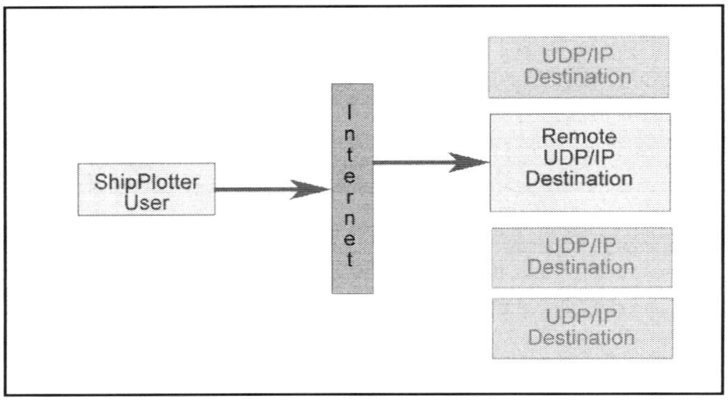

Figure 50 - UDP/IP uploading

To upload and share data using this method the user must first set-up a unique IP address and associated Port number for each destination in the "Options, IO" settings dialog as shown in Figure 51 below.

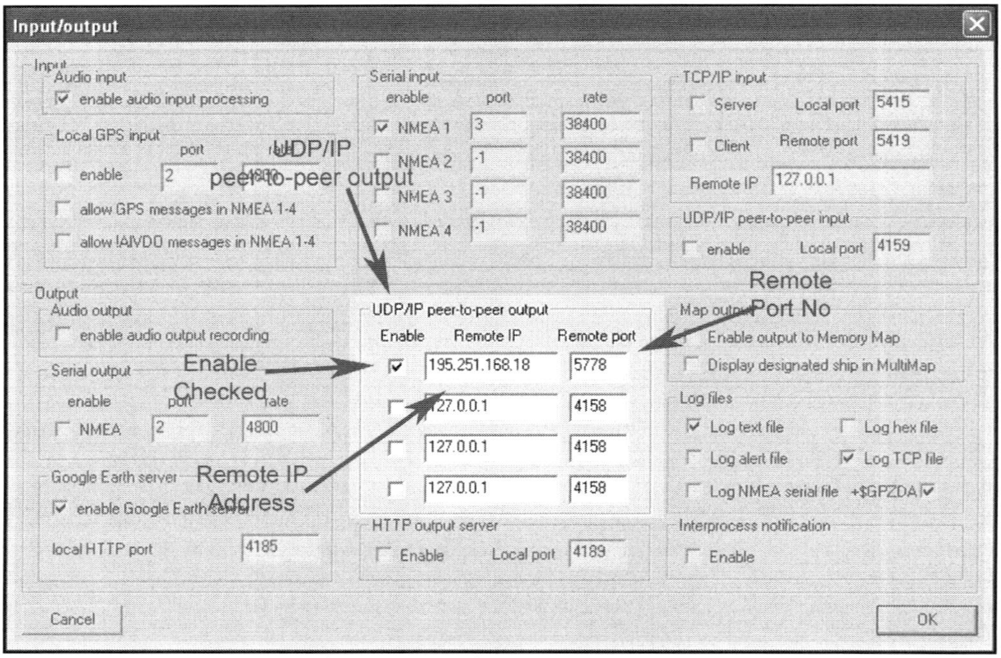

Figure 51 - UDP/IP peer-to-peer output

ShipPlotter as its own Sharing Server

A users individual copy of ShipPlotter can be configured as a sharing server and offer its message data to an unlimited number of other users.

Users wishing to use this feature should consult the ShipPlotter Sharing Tutorial for further information.

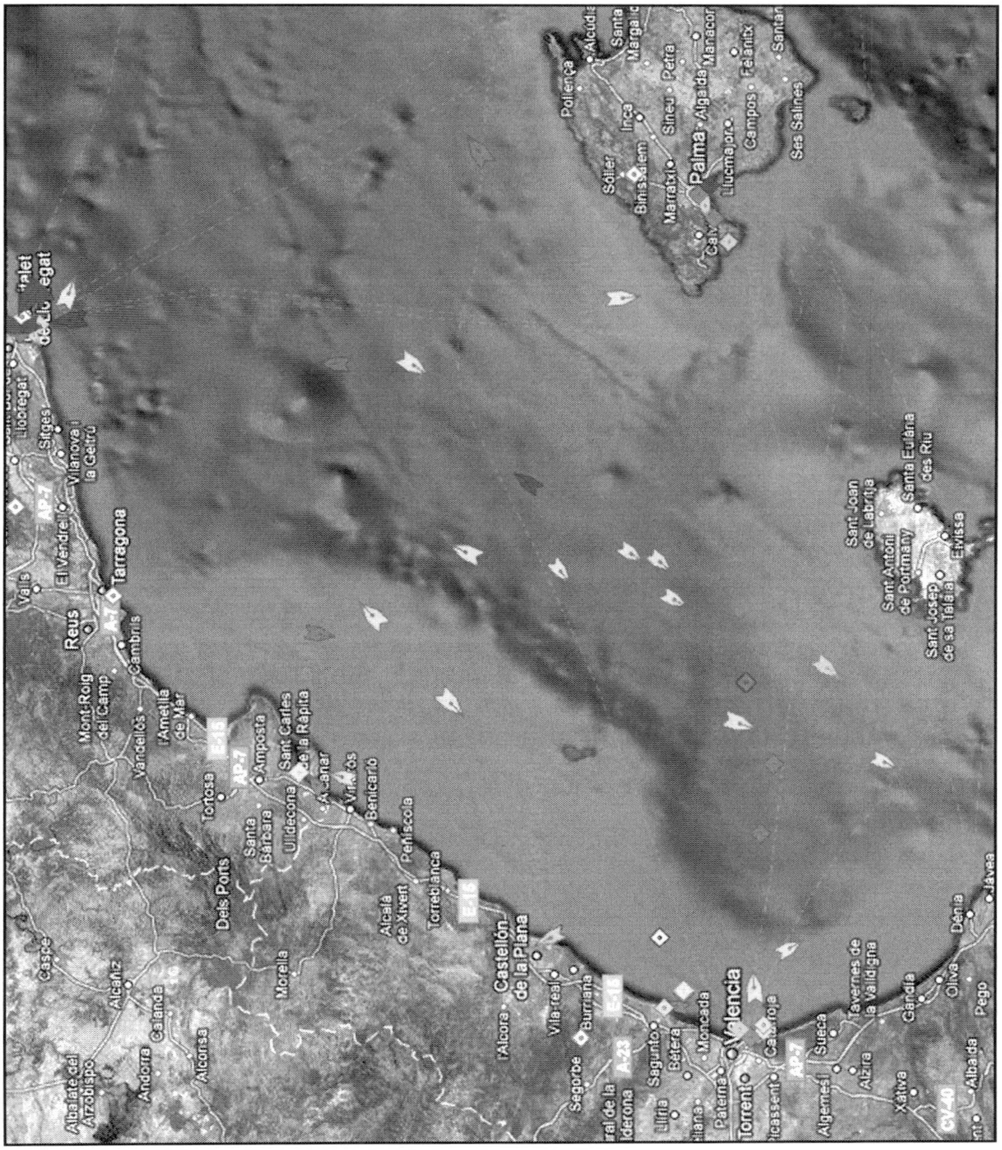

Figure 52 - Marine Traffic Server displaying shared data from a number of sources

Lightship - Calshot Spit
Photo Courtesy and Copyright of Ken Lubi

Chapter 10

Google Maps & Web Pages

As well as displaying the positions of ships and information about them on ShipPlotters own charts and screen display ShipPlotter can transfer the data to other programs such as Google Maps to create custom web page displays.

A number of small computer programs known as scripts have been created and placed in the public domain by other ShipPlotter users and enthusiasts and these can be used to readily incorporate into custom web pages displaying ship information from ShipPlotter.

SP2GM

The first of these scripts is SP2GM (ShipPlotter-to-Google Maps). Running this script will open a copy of Google Maps displayed on a browser web page on the user computer to display ship information derived from a currently running copy of ShipPlotter on the same computer. To use this the user does not need to know anything about web page design or web hosting. The only requirement is for a web browser to be installed on the computer and for the computer to be connected to the Internet. Further details on SP2GM and how to install it will be found in Appendix B.

Source

A Zip file containing the SP2GM script can be downloaded from the Files section of the ShipPlotter Forum at http://groups.yahoo.com/group/shipplotter.

Installation

The downloaded Zip file contains 11 files and these should be extracted to the SPTools folder. "C:\ShipPlotter\SPTools".

SP2GM.html	*SP2GM.lnk*	*SP2GM.txt*
SP2GM.vbs	*BristolChnl GM.lnk*	*Designated.lnk*
divlsc7.gif	*horn0.wav*	*sample-sharecode.dir*
shipplotterlogo_sm.jpg		*shiptrkicon.gif*

To complete the installation create a desktop shortcut with the target entry “C:\Program Files\SPTools\SP2GM.vbs”

Customisation

With a text editor open the “sp2gm.vbs” file to edit the *“USER DEFINITIONS”* section of the script. The first 46 lines of the script are shown below. The user should edit lines 18, 21, 30, 36 & 39 to suit their own installation (these lines are shown in red text for convenience). The users SP share code and the disk drive letter where SP2GM is installed should be specified. Then the full path and file name of the preferred web browser should be selected. Microsoft Internet Explorer is the default and no change will be needed if that is the browser being used. Examples of three other browsers are included as commented lines. Any one of these can be used by removing the “comment apostrophe” and then adding a “comment apostrophe” to comment out the IE browser by adding. A Google Map web page can display map, satellite, or hybrid views. The default with SP2GM is set on Line 36 with “maptyp” where hybrid is the default and map or satellite are alternatives.

```
1 ‘-----------------------------------------’
2 ‘-- SP2GM Version 1.8d November 12, 2007 --’
3 ‘-- Copyright © W. Curt Deegan, 2007 --’
4 ‘-----------------------------------------’
5
6 Option Explicit
7 Dim my_share, my_name, my_drive, quote, dobox, update, stuck, olddsg, quietmode
8 Dim PI, xlat, xlon, xsiz, marklat, marklon, cntrlat, cntrlon, cntrsiz, marknam
9 Dim getshare, locate, range, bearing, minrange, minbearing, mincode, mincnt
10 Dim script, title, ship, browser, web, maptyp, timzon, maxlat, minlat, maxlon, minlon
11 Dim athere, nowCis, toCbe, nowDis, toDbe, nowEis, toEbe
12 Dim MyObject, objFSO, objShell, IE, strPath, strFile, strDirectory, strSoundFile
13 Dim codsearch, namsearch, strsearch, dsgsearch, fixed, browse
14
15 '-- USER DEFINITIONS --'
16
17 '-- SET your SP share code d0'
18 my_share = "a0"
19
20 '-- SET the drive where SP2GM was installed, C: is the default --'
21 my_drive = "C:"
22
23 '-- SET the path and file name of your preferred browser --'
24 '-- For MSIE, browser = "" should be set.
25 '-- To use one of these examples,
26 '-- remove the apostrophe from: 'browser = ...
27 '-- and place an apostrophe before all other browser entries
28 'browser = "" '-- MS IE - leave path and file name blank --'
29 'browser = "C:\Program Files\Netscape\Netscape Browser\
```

```
   netscape.exe"
browser = "C:\Program Files\Mozilla Firefox\firefox.exe"
'browser = "C:\Program Files\Opera\Opera.exe"
'browser = "C:\Program Files\K-Meleon\k-meleon.exe"

'-- SET the default type of Google Map --'
'-- Choices are: map sat hyb
maptyp = "map"

'-- SET the label to be used on the center-of-map marker --'
my_name = "My Home Station Name"

'-- Set True to silence waiting pip --'
quietmode = False

'-- END OF USER DEFINITIONS --'

'-- Program set-up --
```

Running SP2GM

To run the SP2GM script simply double-click the SP2GM.vbs name in the SPTools file folder or click on the desktop shortcut created on installation.

Ship

The first script, SP2GM, is fine for displaying a local Google Map display. However many users have their own web sites and like to publish their received ship data on a public web page Google Map.

The Ship script (now at version nine and titled Ship9.vbs) enables users to create a special unloadable data file in a format that can be recognised by a Google Map Application Programming Interface (API). This file is called "data.xml" and is created in the ShipPlotter Log Files folder. Further details on the Ship9.vbs script and how to install it will be found in Appendix C.

Source

A Zip file containing the Ship9 script can be downloaded from the Files section of the ShipPlotter Forum at http://groups.yahoo.com/group/shipplotter.

Installation

The downloaded Zip file contains 3 files (Ship9.vbs, ReadMe.txt, horn3.wav). These should be extracted to the SPTools folder. “C:\ShipPlotter\SPTools”.

To complete the installation a desktop shortcut with the target entry “C:\Program Files\SPTools\Ship9.vbs”

Customisation

With a text editor open the “Ship9.vbs” file to edit the “USER DEFINITIONS” section of the script. Lines 36-100 of the script are shown below. The user should edit lines 39, 43, 45, 53, 55, 58. 61, 65, 70, 73, 75, 77, 81, 83, 85, 87, 89, 91, 93 & 96 to suit their own installation (these lines are shown in red text for convenience). Lines 406-408 can also be customised as shown.

```
36  '-- BEGIN USER DATA SETTINGS --'
37
38  ' ***SET the two characters of your ShipPlotter share code:
39  my_share = "a0"
40
41  ' ***SET True (vs False) if the home AIS receiver location is to
    be plotted.
42  ' If True, then set name, and local latitude and longitude:
43  my_show = True
44  ' ***SET name of home AIS receiver location:
45  my_name = "My Home Station Name"
46
47  ' ***SET home latitude in decimal degrees for plotting AIS
    receiver location.
48  ' (Should match setting on web page for range/direction
    calculations.)
49  ' (Must use quotes around both values. Use period (.) as decimal
    separator.
50  ' If home location does not plot correctly and comma (,) is your
51  ' usual decimal point, then try using a comma instead of a
    period.)
52  ' Also used if farthest ship is to be calculated and logged:
53  my_lat = "40.3634"   '-- Negative for south --'
54  ' ***SET home longitude:
55  my_lon = "-0.3899" '-- Negative for west --'
56
57  ' ***SET path to where the active ship XML data file is to be
    created:
58  strDirectory = "C:\ShipPlotter\log files\"
59
60  ' ***SET True to restrict ships included to those within SP
    screen lat & lon:
61  useBBox = True
62
```

```
63  ' ***SET True to use range and bearing provided by SP,
64  ' False to compute using coordinates entered in my_lat and my_
    lon:
65  useSPrb = True
66
67  ' ***SET True to compute farthest ship and maintain a log:
68  doRange = True
69  ' ***SET name of farthest ship log file:
70  strLogFile = "farthestlog.txt"
71
72  ' ***SET True to issue range limit alerts, visual and aural:
73  doAlert = False
74  ' ***SET number of nautical miles which when exceeded produces an
    alert:
75  alertRange = 100
76  ' ***SET path to sound file to play on range alert:
77  strSoundFile = "C:\Program Files\COAA\ShipPlotter\horn3.wav"
78
79  ' ***SET True to FTP the active ship XML data file to a web site
    server.
80  ' If True, also set FTP address, user name, password, and
    relative path:
81  doFTP = True
82  ' ***SET FTP address for your web site server:
83  FTPaddr = "http.mysite.com"
84  ' ***SET FTP username for accessing your web site server:
85  UserName = "user_name"
86  ' ***SET FTP password for your server:
87  Password = "user_password"
88  ' ***SET path to subdirectory (folder) on FTP server to put XML
    data file:
89  RelPath = "/mywebsite/public_html/"
90  ' ***SET any extra command to send to ftp server, eg "passive":
91  Ftpextra = "passive"
92  ' ***SET name of XML data file to upload to web site (best to not
    change):
93  strFile = "data.xml"
94
95  ' ***SET True to display processing progress (primarily for
    debugging):
96  ShowSuccess = False
97
98  '-- END USER DATA SETTINGS --'
99
100
```

The user may also wish to change lines 406-408 as shown below to add their own geopgarphical co-ordinates and station name:

```
406 ais = "!!!!!!!Home!!!!" & fixTime() & "!" & my_share & "!L"
407 xml = "<marker lat='40.36' lon='0.39' name='Name of my home
    station' dx='0' dy='0' type='132' ais='" & ais & "'/>"
408 'leave blank
```

Data XML File

Each time the "Ship9.vbs" script is executed a new "data.xml" file is created that looks typically like the example shown below:


```
<markers>
<marker lat='40.61' lon='0.603' name='ALMUDAINA DOS' dx='0' dy='0'
type='40' ais='224596000!9141833!EAEY!S CARLOS DE LA RAPIT!Sep25 13:
00!0.1!180.9!5!100!17!2.5!2009-10-06 12:48:41UTC!dd!A'/>
<marker lat='40.611' lon='0.601' name='MILENIUM' dx='0' dy='0'
type='40' ais='224954000!9221346!EHZN!SAN CARLOS RAPITA!Jan03 14:00!0!
191!5!96!26!4.3!2009-10-06 12:53:45UTC!dd!A'/>
<marker lat='39.567' lon='2.645' name='MRCC PALMA' dx='0' dy='0'
type='101' ais='002241005!0!navaid!....................!...........!0!
0!15!0!0!0!2009-10-06 12:53:47UTC!dd!N'/>
<marker lat='39.874' lon='1.192' name='JESSICA B' dx='-0.011' dy='-
0.02' type='79' ais='304715000!9226384!V2OU8!CEUTA!Oct07 17:00!14.9!21
0!0!133!20!7.5!2009-10-06 12:53:53UTC!dd!A'/>
<marker lat='40.36' lon='0.39' name='Peniscola' dx='0' dy='0'
type='132' ais='!!!!!!!Home!!!!2009-10-06 12:54:02UTC!d0!L'/>
</markers>
```


Each line of data in the "data.xml" file provides the latitude, longitude and name of a ship together with the basic AIS message details and time the message was received.

Once the file has been created then the "Ship9.vbs" script uploads the "data.xml" file to the users web site using a File Transfer Program (FTP).

MoveItFreely FTP

Version 9 of the Ship script use the "MoveItFreely" FTP program in place of any other Windows FTP program that might be installed on the users computer. The author of the Ship script has chosen to use MoveItFreely as it has improved error recovery that will eliminate many problems that might result from ISP or FTP server access problems as well as local PC issues.

The MoveItFreely program can be downloaded without charge from http://www.stdnet.com/products/?category_number=6&subcategory_number=1. The program is self-installing; all that is required is to run the downloaded "MOVEit_Freely_Install.exe" program.

Running Ship9

To run the Ship9 script simply double-click the Ship9.vbs name in the SPTools file folder or click on the desktop shortcut created on installation.

However doing so will only create and upload the latest set of ship information in the "data.xml" file.

What is required however is to update the data on a regular basis and this can easily be achieved using the Windows Schedule program as will be discussed next.

Scheduling

The "Ship9.vbs" script needs to be run regularly every few minutes or so to generate an up- to-date "data.xml" file. This can be done as a Windows "Scheduled Task." To set-up a scheduled the use should first open the Windows Control Panel and then select "Scheduled Tasks".

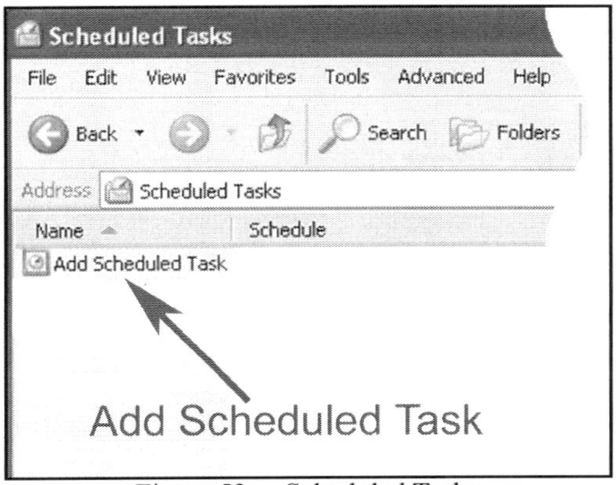

Figure 53a - Scheduled Tasks

1. Click Add Scheduled Task and select next.

2. Click Browse, select the ship9.vbs file and select next.

3. Give a name to the task. For instance "ShipPlotter Update"

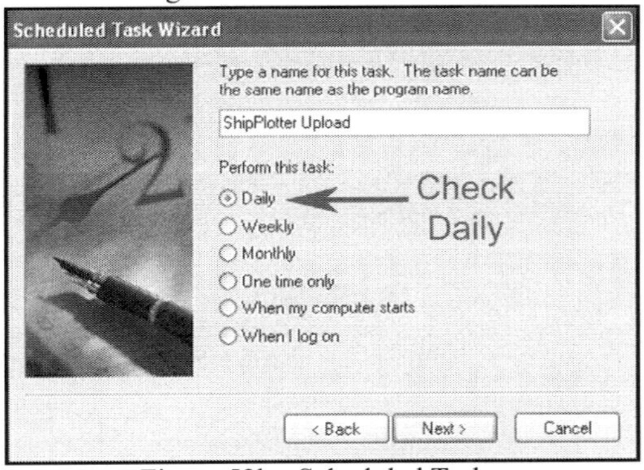

Figure 53b - Scheduled Tasks

4. Check the Daily box in Perform this task.

5. In the next dialog leave the start time and date as today's defaults.

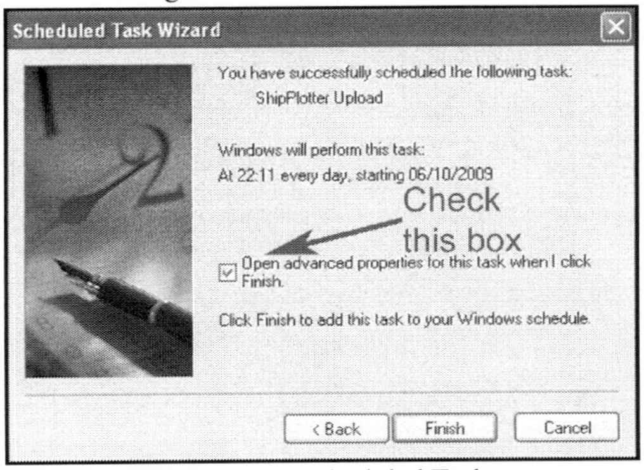

Figure 53c - Scheduled Tasks

6. At the last screen before clicking the "Finish" button check the "Open advanced properties" box.

If the user should forget to select the tick box they can always right click on the ShipPlotter task and then select Properties.

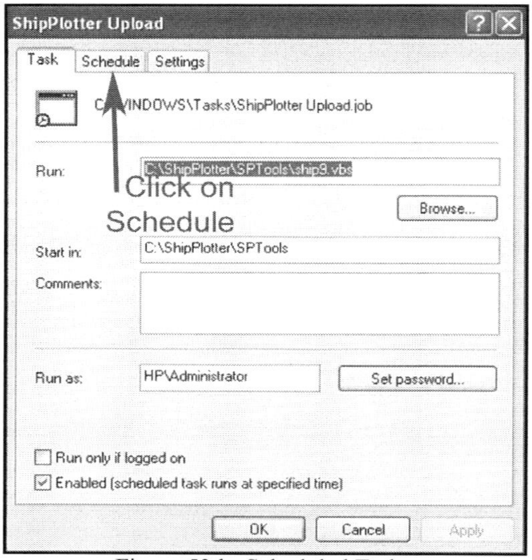

Figure 53d - Scheduled Tasks

7. Select the Schedule tab at the top of the window.

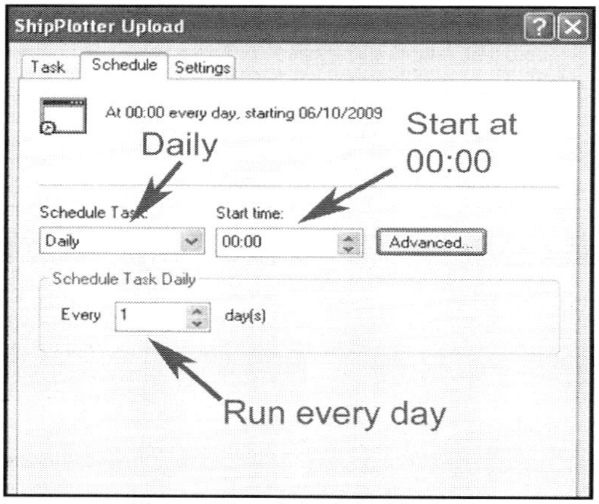

Figure 53e - Scheduled Tasks

8. Schedule Task: Daily

9. Set the Start time to 00:00

10. Schedule the task to run Daily.

11. Click the Advanced button.

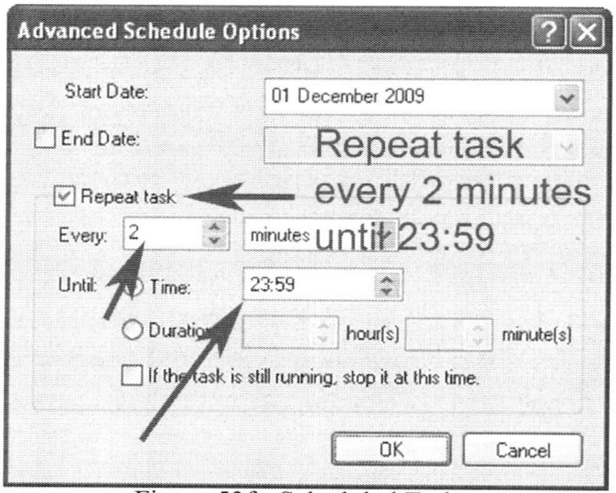

Figure 53f - Scheduled Tasks

12. Check the "Repeat Task box.
14. Set the task to run every "x" minutes.
15. Set the talk to run until 23:59 hours.
16. Click OK to close and save the schedule.

Windows will now run the Ship9.vbs script every 2 minutes automatically each day uploading the data.xml to the user web site.

Google Map Web Page

As stated previously many users have their own web sites and publish their received ship data on a Google Map web page. Examples of such sites are shown below and on the following pages and include:

www.northernbaltic.se/ais/google.html
my.tbaytel.net/va3rom/AIS.htm
www.john-ambler.com/ais/google.html

A Google Map can easily be incorporated into a user web page using some visual basic script and code and deriving live AIS data from the previously described data.xml files.

An example of the code for a complete Google Map web page is shown in Appendix B. This code can be readily modified to customise the page for individual users. The only requirement is for the user to obtain their own Google Map API key from http://www.google.com/apis/maps/.

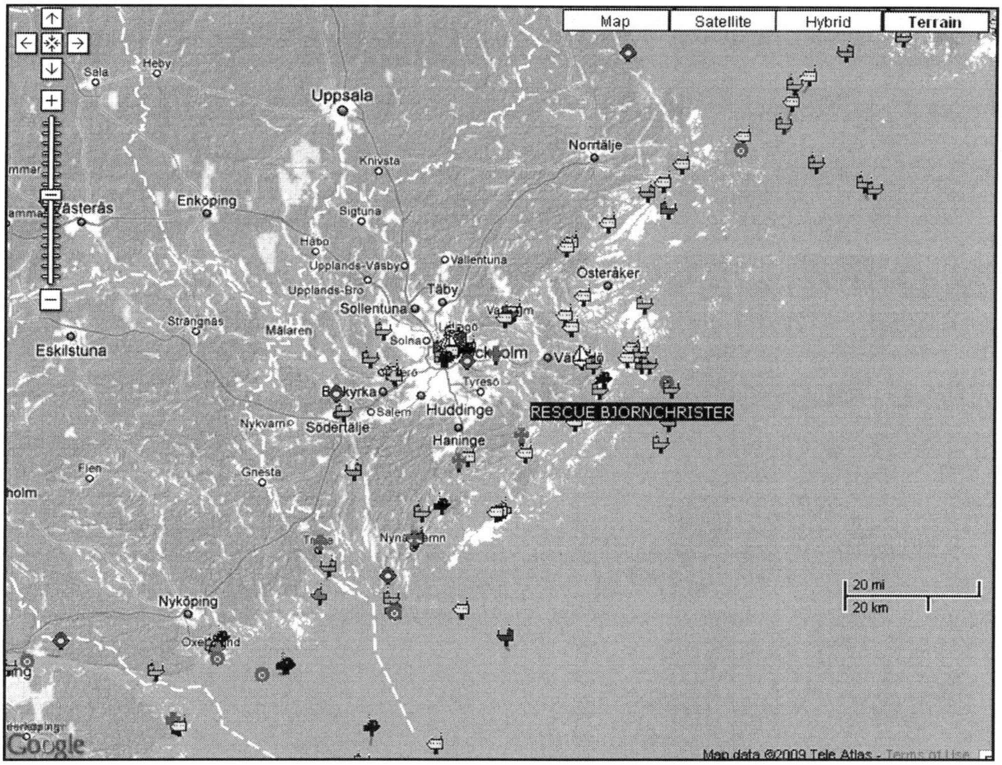

Figure 54 - http://www.northernbaltic.se/ais/google.html
Courtesy Tore Jacobsson

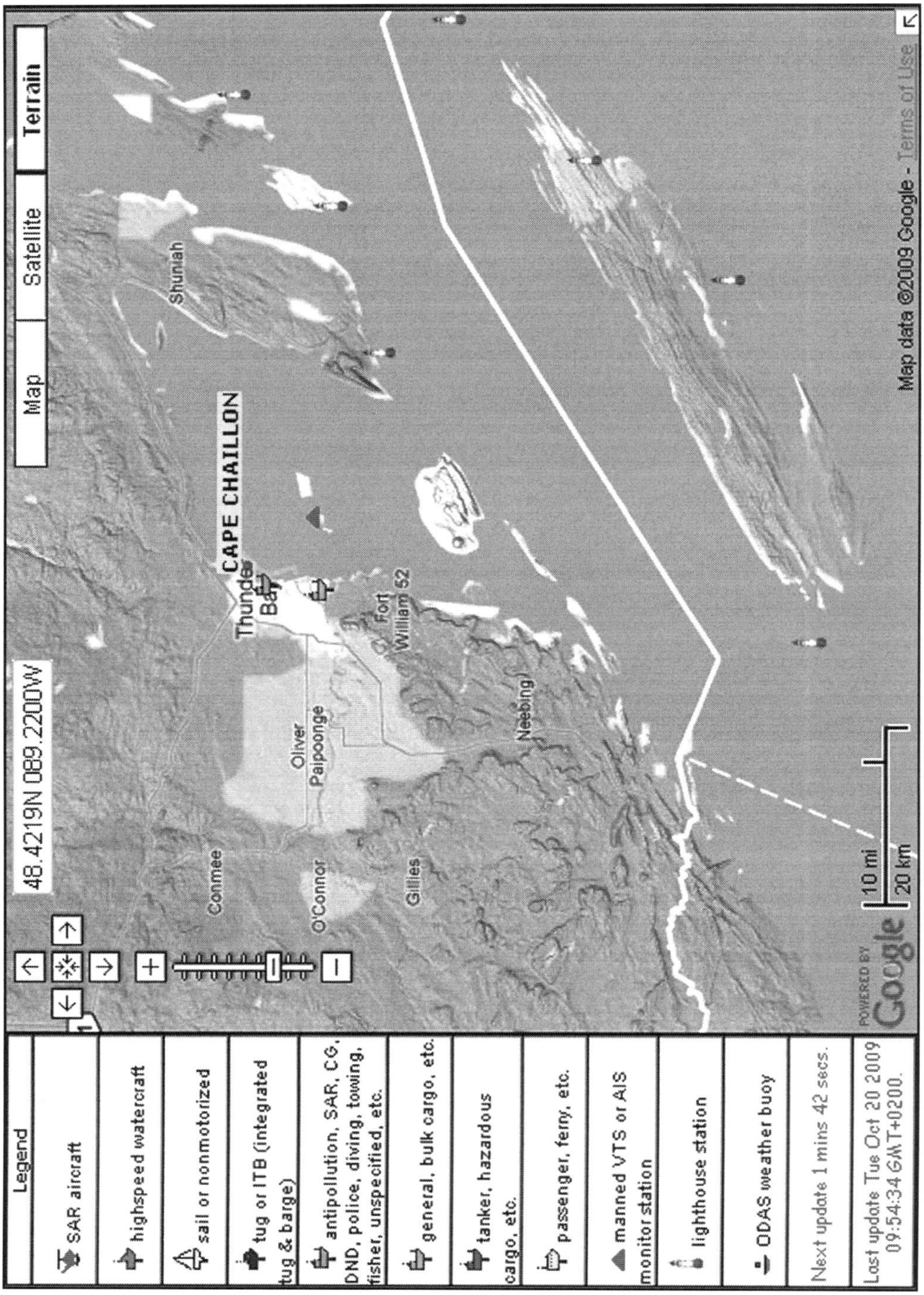

Figure 55 - http://my.tbaytel.net/va3rom/AIS.htm
Courtesy Robert Mazur

Figure 56 - http://www.john-ambler.com/ais/google.html
Courtesy John Ambler

Google Earth Output

If the Google Earth is installed on the users computer then ShipPlotter can display the ships that are currently displayed by ShipPlotter on a Google Earth view. Checking the option and specifying the IP port address in the dialog box as shown in Figure 57 will enable the Google Earth Server.

Google Earth makes use of KML files and for ShipPlotter ships to be displayed a special KML file is required. This file, called 'google_ships.kml', is already provided with ShipPlotter and is installed into the C:\ShipPlotter folder. Having already set-up Google Earth Server as above and making sure that the ShipPlotter program is running and processing data the user should then open the Google Earth program. Then from the Google Earth menu bar select *"File, Open"* to open "C:\ShipPlotter\google_ships.kml".

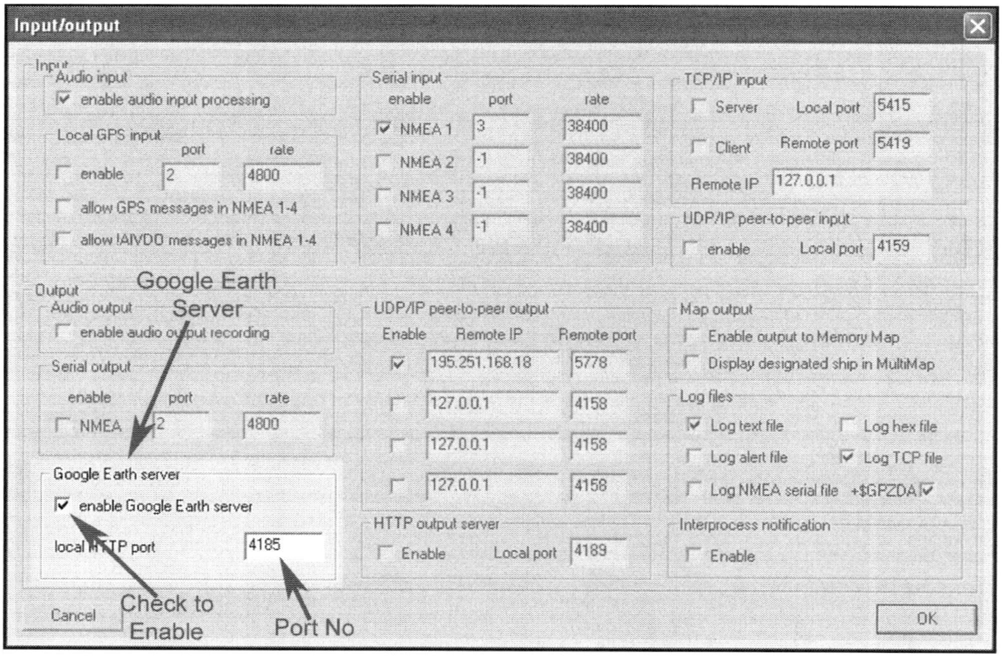

Figure 57 - Google Earth Set-up

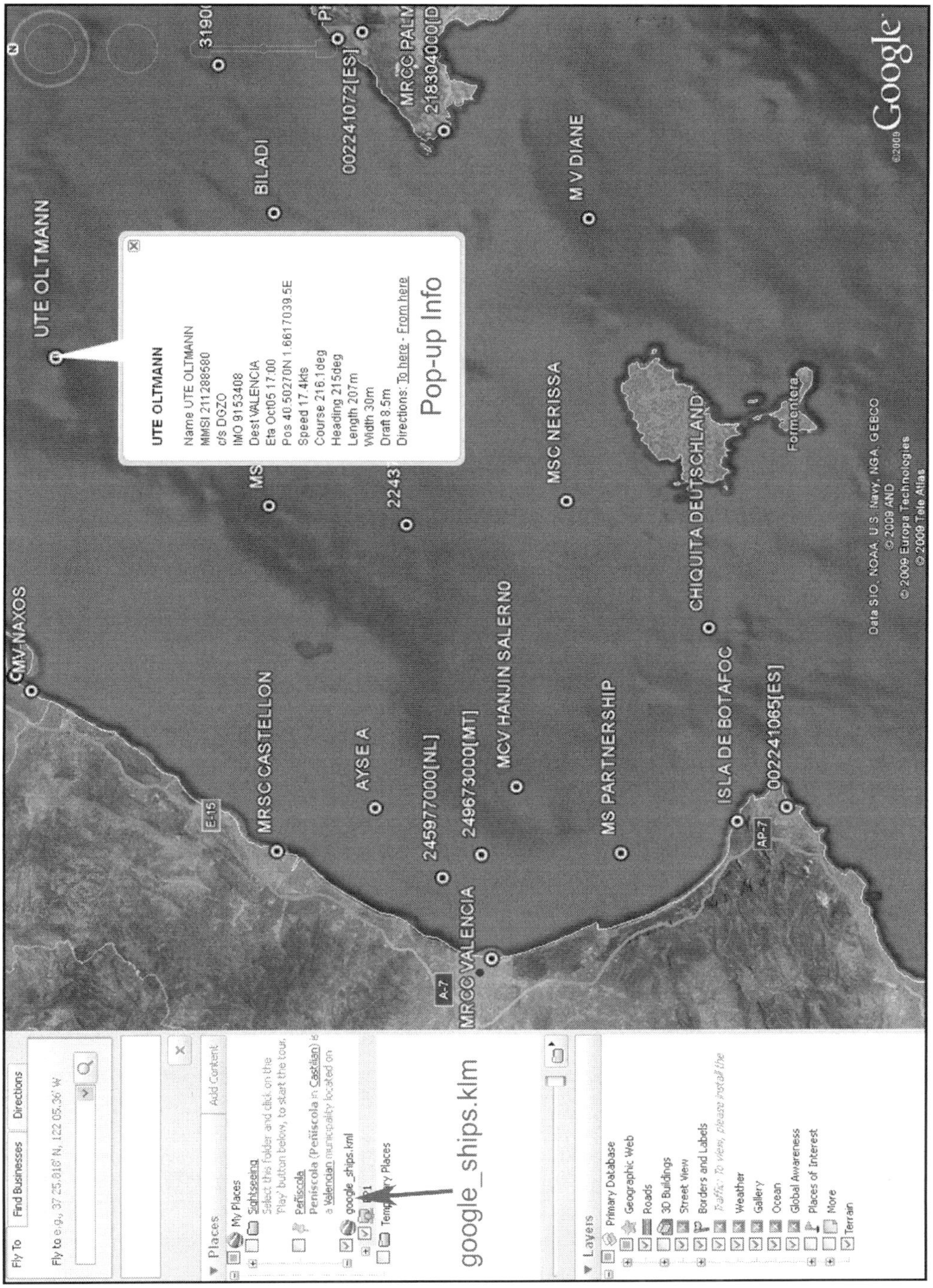

Figure 58 - Google Earth Display with Ships

Passenger - Queen Mary 2
Photo Courtesy and Copyright of Daniel Ferro

Chapter 11
User Support

FAQs

Q. Why is there is no signal?

A. *If there is no signal at all, use the "Options, "Audio" to select the Line-In option and to set the volume control for the Line-In (and the overall volume control if there is one), to obtain a signal. If there is no Line-In option (sometimes called Aux), in the volume control dialog click on "Options, Properties, Recording", check Line-In and click on OK.*

Q. Why does sharing not work?

A. *If sharing cannot be enabled then either the software is not registered or the settings in the "Options, Demodulator" dialog are not correct. ShipPlotter will not allow users to share their data if they have disabled the checksum checking in that dialog. This is to avoid distributing corrupt messages to other users.*

Q. What is my sharing ID?

A. *Use the "Help, About" menu option to display the version of ShipPlotter installed on the computer. It will show a line like 'ShipPlotter version 12.4.5 - Registered {1234}d0'. The last two characters, d0 in the example above, are the users sharing ID. Once this is found it should be entered in the Sharers database at http://groups.yahoo.com/group/shipplotter/database.*

Q. Is it important to ensure that my computer has the correct time?

A. *If the computers time is not accurate then the sharing of data may be compromised thus it is very important to synchronise a computers clock with a national standard. This can be achieved automatically by using a time setting and maintenance software program such as NTP (Network Time Protocol) from http://www.meinberg.de/english/sw/ntp.htm.*

Q. How can one check that signals are being received?

A. *The I/O Monitor provides a graphical indicator of data received and sent.*

Q. Why is shared data is not being downloaded?

A. *The Audio Input "enable audio input processing" box in the I/O dialog must be checked to process data downloaded from a sharing server.*

Q. There is a note displayed in the ShipPlotter Status Bar "No messages received from *ship name*, why?

A. *This is because the user has designated a ship at some time. Double clicking on it in the chart display designates a ship. The data for the designated ship appears in the status bar of ShipPlotter. The user can unset the designated status by double clicking it a second time or by designating a new ship.*

Q. How can ShipPlotter start processing data as soon as the program is loaded?

A. *Command line switches can be used with ShipPlotter. For example the command line "C:\ShipPlotter\ShipPlotter.exe /start" will load the program and start processing data immediately.*

Q. How can ShipPlotter change the chart views without loosing the currently displayed data?

A. *By using the Quick Charts options. As Quick Charts are saved with a configuration changing from one Quick Chart view to another retains all the current data.*

Q. Can I use ShipPlotter on more than one computer?

A. *A registered user can use ShipPlotter on more than one computer but they will need a new licence key for each different PC.*

Q. How can a user obtain a new licence key?

A. *By visiting the main ShipPlotter site at http://www.coaa.co.uk/COAAreg.htm. If the user already has a licence then they simply have to enter in their details to generate a new serial number.*

Q. How can a user obtain a new licence key?

A. *By visiting the main ShipPlotter site at http://www.coaa.co.uk/COAAreg.htm Entering a users current details will automatically generate a new key.*

User Forum & Support

There is a well-established ShipPlotter User Group that has its own web forum at http://groups.yahoo.com/group/shipplotter.

Support can also be obtained from COAA the authors of the ShipPlotter software program by email info@coaa.co.uk.

Other Software

There are a number of useful documents and software add-ons that can be used in conjunction with ShipPlotter including amongst others a Technical manual for the SR162 AIS Receiver, a number of calibrated charts for different locations around the world, an add-on to check a users sharing status and programs for analysing received AIS data.

These can be found in the Files section of the ShipPlotter User Group at http://groups.yahoo.com/group/shipplotter/files.

This book

Copies of the scripts and google.html as well as Electronic Updates to the book will be provided free of charge to bona fide purchasers of the book. Comments or suggestions on this book should be addressed by email to the author at shipplotter@kenandglen.com or by visiting his web pages at http://www.kenandglen.com.

Appendix A - AIS Types

With the detailed AIS information the user will see a type description (e.g. Cargo) and a type number (70). The type number is what AIS actually broadcast and this is converted to a description using the following categories:

The first digit represents the general category

1. Reserved
2. Wing In Ground
3. Special category
4. Hi Speed Craft
5. Special category
6. Passenger
7. Cargo
8. Tanker
9. Other

While the second digit gives some information as to the cargo

1. Major Hazard (Haz A)
2. Hazard (Haz B)
3. Minor Hazard (Haz C)
4. Recognisable Hazard (Haz D)
5. undefined
6. undefined
7. undefined
8. undefined
9. undefined

Special categories that start with 3

1. Fishing
2. Towing
3. Towing
4. Dredger
5. Dive Vessel
6. Military Operations
7. Sailing
8. Pleasure Craft
9. Reserved

Special categories that start with 5

1. Pilot Vessel
2. Search and Rescue
3. Tugs
4. Port Tenders
5. Anti-pollution
6. Law enforcement
7. Local Vessel
8. Local Vessel
9. Medical Transport
10. Special craft

As can be seen the types definitions are a bit vague and have gaps in. Is a RORO ferry a passenger ship carrying cargo or a cargo ship carrying passengers? However it is not uncommon for the operators of AIS on board ship to not always be as careful as they should be in filling out the numbers so occasionally one might see a ship with for example type 89. This probably means a tanker of some sort but there is no definition as to exactly what the 9 means.

Values over 100

100	Default Navaid	
101	Reference point	
102	RACON	
103	Offshore Structure	
104	Spare	
105	Light	without sectors
106	Light	with sectors
107	Leading Light	Front
108	Leading Light	Rear
109	Beacon	Cardinal N
110	Beacon,	Cardinal E
111	Beacon	Cardinal S
112	Beacon	Cardinal W
113	Beacon	Port hand
114	Beacon	Starboard hand
115	Beacon	Preferred Channel port hand
116	Beacon	Preferred Channel starboard hand
117	Beacon	Isolated danger
118	Beacon	Safe water
119	Beacon	Special mark
120	Cardinal Mark N	
121	Cardinal Mark E	
122	Cardinal Mark S	
123	Cardinal Mark W	
124	Port hand Mark	
125	Starboard hand Mark	
126	Preferred Channel	Port hand
127	Preferred Channel	Starboard hand
128	Isolated danger	
129	Safe Water	
130	Manned VTS / Special Mark	
131	Light Vessel / LANBY	

Appendix B - Google HTML

Appendix B contains the code for a simple HTML web page to display a Google Mapwith ship information derived from a currently running copy of ShipPlotter.

Source

The code listed is a based on Curt Deegans original "google.html" file that is part of the "vb4ship.zip" package that can be downloaded from the Files section of the ShipPlotter Forum at http://groups.yahoo.com/group/shipplotter. It is recommended that users first download that file and then modify to the code listed below rather than typing in all the code from here.

Customisation

The only customisation required is to change the Google Maps API key in line 53 of the listed code. The user should signup to get their own API key from http://www.google.com/apis/maps/. The key will look like a long strings of letters and characters should be used to replace the key=ABQIAAAAvpcmg……...

The entry (new GLatLng(40.36, 1.1), 9) on line 62 must be also be changed to the users home station co-ordinates. Note the number 9 in that entry is the Google Map zoom level and it is not recommended to change that value to start off with. The Home Statio co-ordinates must also be entered in lines 235 &236 to enable the range and direction of ships to be correctly calculated.

The user may also wish to change the variables 'refreshInterval' and 'numrefresh' which set how often the page is refreshed (in seconds) and how many refreshes are allowed before the page times out. These can be found in lines 586-590.

Code

```
1 <!DOCTYPE HTML PUBLIC “-//W3C//DTD HTML 4.01 Transitional//EN”
2 “http://www.w3.org/TR/html4/loose.dtd”>
3 <html>
4 <head>
5 <meta http-equiv=”Content-Type” content=”text/html; charset=iso-
   8859-1”>
6 <title>My ShipPlotter Google Map</title>
7 <style type=”text/css” media=”screen”>
8 v\:* {
9 behavior:url(#default#VML);
10 }
11 .gpopup {
12 width:190px;
13 }
14 .gpopup table {
15 padding:0px;
16 border-collapse:collapse;
17 border:0;
18 font-family: Verdana, Arial, Helvetica, sans-serif, “MS sans
   serif”;
19 font-size: 9px;
20 text-align:left;
21 color: black;
22 background: white;
```

```
}
#tooltip span {
color:black;
background:white;
font-weight:bold;
font-size: 10pt;
font-family: Verdana, Arial, Helvetica, sans-serif, "MS sans
  serif";
white-space: nowrap;
}
.style2 {color: #CCCCCC}
.style11 {
color: #FFFFFF;
font-size: x-small;
}
.style13 {color: #CCCCCC; font-size: x-small; }
}
.style20 {font-size: x-small}
.style21 {
font-size: small;
font-weight: bold;
}
body {
margin-top: 0px;
}
.style24 {color: #FFFFFF; font-size: x-small; font-weight: bold; }
.style26 {color: #FFFFFF; font-size: small; }
.style27 {color: #FF0000; font-size: x-small; }
</style>
<!-- Get an API key from http://www.google.com/apis/maps/ -->
<meta http-equiv="Content-Language" content="en-gb">
<script src="http://maps.google.com/maps?file=api&amp;v=2&amp;key=A
  BQIAAAAvpcmgFjMSb-UIGpaRv1v1xQ8pXTcUDlOEMLv_QWDfluaaQZdqxS0cCCk737
  f2SxInb09r2rRgZtIWg"
type="text/javascript"></script>
<script type="text/javascript">
//<![CDATA[
function loadup() {
map = new GMap2(document.getElementById("map"));
map.addControl(new GLargeMapControl());
map.addControl(new GMapTypeControl());
map.addControl(new GScaleControl());
map.setCenter(new GLatLng(40.36, 1.1), 9); // Change these to your
  own part of the world
map.addMapType(G_PHYSICAL_MAP);
// ====== set up marker mouseover tooltip div ======
tooltip = document.createElement("div");
map.getPane(G_MAP_FLOAT_PANE).appendChild(tooltip);
tooltip.style.visibility="hidden";

icon_k = myicon("http://www.kenandglen.com/images/mm_20_
  black.png", 0);
icon_b = myicon("http://www.kenandglen.com/images/mm_20_blue.png",
```

```
    0);
icon_a = myicon(“http://www.kenandglen.com/images/
    lighthouse1.png”, 0);
icon_g = myicon(“http://www.kenandglen.com/images/mm_20_
    green.png”, 0);
icon_r = myicon(“http://www.kenandglen.com/images/lifeboat2.png”,
    0);
icon_w = myicon(“http://www.kenandglen.com/images/mm_20_
    white.png”, 0);
icon_y = myicon(“http://www.kenandglen.com/images/mm_20_
    yellow.png”, 0);
icon_m = myicon(“http://www.kenandglen.com/images/mm_20_
    purple.png”, 0);
icon_o = myicon(“http://www.kenandglen.com/images/mm_20_
    orange.png”, 0);
icon_n = myicon(“http://www.kenandglen.com/images/mm_20_
    brown.png”, 0);
icon_t = myicon(“http://www.kenandglen.com/images/mm_20_
    turkos.png”, 0);
icon_h = myicon(“http://www.kenandglen.com/images/heli4.png”, 2);
icon_home = myicon(“http://www.kenandglen.com/images/marker.png”,
    0); // Added to display Home AIS
icon_track = myicon(“shiptrkicon.gif”, 1); // Added to display
    replay tracks

autoUpdate();
}
function ais_type(id) {
var type1 = new Array(‘unspecified’, ‘Reserved’, ‘Wing In Grnd’,
    ‘’, ‘Hi Spd Crft’, ‘’, ‘Passenger’, ‘Cargo’, ‘Tanker’, ‘Other’);
var type2 = new Array(‘’, // All ships of this type
‘Haz A’, /* Major Hazard */
‘Haz B’, /* Hazard */
‘Haz C’, /* Minor Hazard */
‘Haz D’, /* Recogniseable Hazard */
‘’,
‘’,
‘’,
‘’,
‘’ // No additional information
);
type3 = new Array(‘Fishing’, ‘Towing’, ‘Towing’, ‘Dredger’, ‘Dive
    Vessel’, ‘Military Ops’, ‘Sail’, ‘Pleasure Craft’, ‘Reserved’,
    ‘Reserved’);
 type5 = new Array(‘Pilot Vessel’, ‘SAR’, ‘Tug’, ‘Port Tender’,
    ‘Anti-polution’, ‘Law enforce’, ‘Local Vessel’, ‘Local Vessel’,
    ‘Medical Trans’, ‘Special craft’);
 navaid = new Array(‘Default Navaid’,
‘Reference point’,
‘RACON’,
‘Off Shore Structure’,
‘Spare’,
‘Light, without sectors’,
```

```
‘Light, with sectors’,
‘Leading Light Front’,
‘Leading Light Rear’,
‘Beacon, Cardinal N’,
‘Beacon, Cardinal E’,
‘Beacon, Cardinal S’,
‘Beacon, Cardinal W’,
‘Beacon, Port hand’,
‘Beacon, Starboard hand’,
‘Beacon, Preferred Channel port hand’,
‘Beacon, Preferred Channel starboard hand’,
‘Beacon, Isolated danger’,
‘Beacon, Safe water’,
‘Beacon, Special mark’,
‘Cardinal Mark N’,
‘Cardinal Mark E’,
‘Cardinal Mark S’,
‘Cardinal Mark W’,
‘Port hand Mark’,
‘Starboard hand Mark’,
‘Preferred Channel Port hand’,
‘Preferred Channel Starboard hand’,
‘Isolated danger’,
‘Safe Water’,
‘Manned VTS’, // Special Mark
‘Light Vessel / LANBY’,
‘Receiving station’); // Added to show unique icon for Home AIS
type = Math.floor(id / 10);
rem = id % 10;
result = ‘’;
if (type == 3)
result = type3[rem];
else if (type == 5)
result = type5[rem];
else if (id >= 100) {
result = navaid[id-100];
} else {
result = type1[type];
if (id < 100 && type2[rem] != ‘’)
result = result +’ ‘+type2[rem];
}
return result;
}
function getstatus(status) {
switch (parseInt(status)) {
case 0: return ‘Underway’;
break;
case 1: return ‘At Anchor’;
break;
case 2: return ‘Not Under Command’;
break;
case 3: return ‘Restricted Maneuverability’;
break;
```

```
case 4: return ‘Constrained by Her Draught’;
break;
case 5: return ‘Moored’;
break;
case 6: return ‘Aground’;
break;
case 7: return ‘Engaged in Fishing’;
break;
case 8: return ‘Underway by Sail’;
break;
case 14: return ‘’;
break;
case 15: return ‘Default (15)’;
break;
default: return status;
break;
}
}
function getaisclass(aisclass) {
switch (aisclass) {
case “A”: return ‘AIS Class A’;
break;
case “B”: return ‘AIS Class B’;
break;
case “L”: return ‘Local GPS’;
break;
case “N”: return ‘Navigation Aid’;
break;
case “P”: return ‘SAR Aircraft’;
break;
case “S”: return ‘AIS Base Station’;
break;
case “?”: return ‘Unkown’;
break;
case “ “: return ‘No Data’;
break;
default: return aisclass;
break;
}
}
function Left(str, n)
{
if (n <= 0)
return “”;
else if (n > String(str).length)
return str;
else
return String(str).substring(0,n);
}
function makeinfo(Lat, Lon, name, type, aisstr, i) {
var ais = aisstr.split(‘!’);
var mmsi = ais[0];
var imo = ais[1];
```

```
var call = ais[2];
var dest = ais[3];
var eta = ais[4];
var speed = ais[5];
var course= ais[6];
var status= getstatus(ais[7]);
var length= ais[8];
var width = ais[9];
var draft = ais[10];
var mtime = new Date(ais[11]);
var mtime= Left(mtime, 24)+"Local";
var shareid = ais[12]; // Added to show share code in info balloon
var aisclass = ais[13]; // Added to show AIS class in info balloon

// ***SET home coordinates for range and direction calculations.
var my_lat = 40.3634;
var my_lon = 0.3899;
// Convert Lat/Long to NSEW versus +- for display in info balloon
with (Math) {
if (Lat < 0) {
var latstr = abs(Lat);
latstr = 'S'+latstr;
} else {
latstr = 'N'+Lat;
}
if (Lon < 0) {
var lonstr = abs(Lon);
lonstr = 'W'+lonstr;
} else {
lonstr = 'E'+Lon;
}
// Calculate the RANGE from home location.
var la1 = my_lat * PI / 180;
var lo1 = my_lon * PI / 180;
var la2 = Lat * PI / 180;
var lo2 = Lon * PI / 180;
var x = (sin(la2) * sin(la1)) + (cos(la2) * cos(la1) * cos(lo2-lo1));
var range = round((((PI / 2) - atan(x / sqrt(-x * x + 1))) * 60 * 180 / PI) * 10) / 10;
// Calculate Initial Bearing of ship from home location.
var y = sin(lo2 - lo1) * cos(la2);
x = cos(la1) * sin(la2) - sin(la1) * cos(la2) * cos(lo2 - lo1);
var bearing = round(((atan2(y, x) * 180 / PI) + 360) % 360);

// Find farthest ship.
if (range > maxrange) {
maxrange = range;
maxname = name;
maxmmsi = mmsi;
maxlat = Lat;
```

```
maxlon = Lon;
maxbearing = bearing;
maxindex = i;
}
}
if ((type < 100) || (type == 133)) {
callstr = ‘<tr><td><b>Callsign:</b></td><td>’+call+’</td></tr>\n’;
deststr = ‘<tr><td><b>Destination:</b></td><td>’+dest+’</td></tr>\n’;
etastr = ‘<tr><td><b>ETA:</b></td><td>’+eta+’</td></tr>\n’;
} else {
callstr = ‘’;
deststr = ‘’;
etastr = ‘’;
}
if (type < 100) {
spdstr = ‘<tr><td><b>Speed&Dir:</b></td><td>’+speed+’ kts / ‘+course+’<sup>o</sup></td></tr>\n’;
sizestr = ‘<tr><td><b>Size:</b></td><td>’+length+’m x ‘+width+’m x ‘+draft+’m</td></tr>\n’;
} else {
spdstr = ‘’;
sizestr = ‘’;
}
if (mmsi != ‘’) {
mmsistr = ‘<tr><td><b>MMSI:</b></td><td>’+mmsi+’</td></tr>\n’;
} else {
mmsistr = ‘’;
}
if (imo > 0) {
imostr = ‘<tr><td><b>IMO:</b></td><td>’+imo+’</td></tr>\n’;
} else {
imostr = ‘’;
}
if (mtime != ‘’) {
recstr = ‘<tr><td><b>Received:</b></td><td>’+mtime+’</td></tr>\n’;
} else {
recstr = ‘’;
}
if (status != ‘’) {
statusstr = ‘<tr><td><b>Status:</b></td><td>’+status+’</td></tr>\n’;
} else {
statusstr = ‘’;
}
// Following entries added to show Share Code, Range & Bearing, and AIS Class
if (shareid != ‘’) {
sharestr = ‘<tr><td><b>Share Code:</b></td><td> ‘+shareid+’</td></tr>\n’;
} else {
```

```
sharestr = ‘’;
}
if (range != ‘’) {
rangestr = ‘<tr><td><b>Range&Bear:</b>  </td><td>’+range+’nm / ‘+bearing+’<sup>o</sup>T</td></tr>\n’;
} else {
rangestr = ‘’;
}
if (aisclass != ‘’) {
classstr = getaisclass(aisclass)
classstr = ‘<tr><td><b>AIS Class:</b></td><td>’+classstr+’</td></tr>\n’;
} else {
classstr = ‘’;
}
var img = “”;
if (imo > 0) {
img = ‘<a href=”http://www.marinetraffic.com/Ais/showphoto.aspx?mmsi=’+mmsi+’”><img width=”200” style=”margin-right:5px” src=”http://www.marinetraffic.com/Ais/showphoto.aspx?mmsi=’+mmsi+’”>’;
}
return ‘<div class=”gpopup”><table><tr><td><table cellpadding=”0” cellspacing=”0”>\n’+
‘<tr><td><b>Name:</b></td><td>’+name+’</a></td></tr>\n’+
callstr+
mmsistr+
imostr+
statusstr+
deststr+
etastr+
‘<tr><td><b>Type:</b></td><td>’+ais_type(type)+’</td></tr>\n’+
sizestr+
spdstr+
‘<tr><td><b>Location:</b></td><td>’+latstr+’<sup>o</sup> / ‘+lonstr+’<sup>o</sup></td></tr>\n’+
rangestr+
classstr+
‘<tr><td colspan=”2”> </td></tr>\n’+
recstr+
sharestr+
‘</table></td><td>’+img+’</td></tr></table></div>\n’;
}
function myicon(image, i) {
icon = new GIcon();
if (i == 1) {
icon.iconSize = new GSize(4, 4);
icon.iconAnchor = new GPoint(2, 4);
icon.shadow = “”;
icon.shadowSize = new GSize(0, 0);
}
if (i == 2) {
icon.iconSize = new GSize(40, 40);
icon.iconAnchor = new GPoint(20, 40);
```

```
icon.shadow = “”;
icon.shadowSize = new GSize(0, 0);
}
else {
icon.iconSize = new GSize(12, 20);
icon.iconAnchor = new GPoint(6, 20);
icon.shadow = “http://www.kenandglen.com/images/mm_20_shadow.png”;
icon.shadowSize = new GSize(22, 20);
}
icon.infoWindowAnchor = new GPoint(5, 1);
icon.image = image;
return icon;
}
function addVector(from, deltax, deltay) {
if (deltax == 0 && deltay == 0) return null;
var to = new GLatLng(from.lat() + deltay, from.lng() + deltax)
var polyline = new GPolyline([from, to], “#00ff00”, 2);
return polyline;
}
function get_colour(type) {
if (type == 14)
return icon_h;
if (type == 30)
return icon_o;
if (type == 31)
return icon_m;
if (type == 32)
return icon_m;
if (type == 33)
return icon_g;
if (type == 34)
return icon_n;
if (type == 35)
return icon_o; // “orange”
if (type == 36)
return icon_b;
if (type == 37)
return icon_b; // “blue”
if (type == 38)
return icon_w;
if (type == 39)
return icon_w;
if (type == 50)
return icon_m;
if (type == 51)
return icon_r;
if (type == 52)
return icon_m; // “magenta”
if (type == 53)
return icon_m;
if (type == 54)
return icon_o;
```

```
if (type == 55)
return icon_o;
if (type == 56)
return icon_w;
if (type == 57)
return icon_w;
if (type == 58)
return icon_r;
if (type == 59)
return icon_r;
if (type == 132)
return icon_home; // Added to display Home AIS
if (type == 133)
return icon_track; // Added to display replay tracks
if (type > 99)
return icon_a; //”gray”
var itype = Math.floor(type/10);
switch (itype) {
case 7:
ball = icon_y; // “yellow”;
break;
case 8:
ball = icon_k; // “black”;
break;
case 4:
ball = icon_t; // “blue”;
break;
case 6:
ball = icon_b; // “blue”;
break;
default:
ball = icon_w; // “white”;
break;
}
return ball;
}
function createMarker(point, name, html, iconz) {
var marker = new GMarker(point, iconz);
marker.tooltip = ‘<div id=”tooltip”><span>’+name+’</span></div>’;
GEvent.addListener(marker, “click”, function() {
marker.openInfoWindowHtml(html);
});
map.addOverlay(marker);
GEvent.addListener(marker,”mouseover”, function() {
showTooltip(marker);
});
GEvent.addListener(marker,”mouseout”, function() {
tooltip.style.visibility=”hidden”
});
return marker;
}
function autoUpdate() {
GDownloadUrl(“data.xml?”+Math.random(), function(data,
```

```
responseCode) {
var xmlDoc = GXml.parse(data);
if (!xmlDoc.documentElement) return;
var xmlmarker = xmlDoc.documentElement.getElementsByTagName(“mar
ker”);
var movehtml = ‘’;
var dockhtml = ‘’;
var sepmove = ‘’;
var sepdock = ‘’;
var movecnt = 0;
var dockcnt = 0;
if (!gmarkers)
gmarkers = new Array();
if (!htmls)
htmls = new Array();
if (!vectors)
vectors = new Array();
var Lat, Lon, name, deltax, deltay, type, ais, html, ball, icon;
for (var i = 0; i < xmlmarker.length; i++) {
Lat = parseFloat(xmlmarker[i].getAttribute(“lat”));
Lon = parseFloat(xmlmarker[i].getAttribute(“lon”));
name = xmlmarker[i].getAttribute(“name”);
deltax = parseFloat(xmlmarker[i].getAttribute(“dx”));
deltay = parseFloat(xmlmarker[i].getAttribute(“dy”));
type = parseInt(xmlmarker[i].getAttribute(“type”));
htmls[i] = makeinfo(Lat, Lon, name, type, xmlmarker[i].getAttribu
te(“ais”), i);
icon = get_colour(type);
if (gmarkers[i]) {
GEvent.clearListeners(gmarkers[i], “click”);
GEvent.clearListeners(gmarkers[i], “mouseover”);
GEvent.clearListeners(gmarkers[i], “mouseout”);
map.removeOverlay(gmarkers[i]);
}
var point = new GLatLng(Lat, Lon);
gmarkers[i] = createMarker(point, name, htmls[i], icon);

if (vectors[i]) map.removeOverlay(vectors[i]);
vectors[i] = addVector(point, deltax, deltay);
if (vectors[i]) map.addOverlay(vectors[i]);
html = “<a href=’javascript:findship(“+i+”)’ onmouseover=’mymouseo
ver(“+i+”)’ onmouseout=’mymouseout()’>”+name+”</a>”;
if (Math.abs(deltax) > 0 || Math.abs(deltay) > 0) {
movehtml += sepmove + html;
sepmove = “, “; movecnt += 1;
} else {
dockhtml += sepdock + html;
sepdock = “, “; dockcnt += 1;
}
}
var shipmove = document.getElementById(“shipmove”);
if (shipmove)
shipmove.innerHTML = ‘<h3>Ships Underway [‘ + movecnt + ‘]: </
```

```
h3><font size="-1">' + movehtml + '</font>.';
var shipdock = document.getElementById("shipdocked");
if (shipdock)
shipdock.innerHTML = '<h3>Ships Docked, Anchored or Fixed [' +
dockcnt + ']: </h3><font size="-1">' + dockhtml + '</font>.';
var shiptime = document.getElementById("shiptime");
if (shiptime) {
var currentTime = new Date();
// See eg http://www.tizag.com/javascriptT/javascriptdate.php for
// fancier time formatting
shiptime.innerHTML = "Ship positions last updated: " +
currentTime + '.';
}
html = "<a href='javascript:findship("+maxindex+")' onmouseover='m
ymouseover("+maxindex+")' onmouseout='mymouseout()'>"+maxname+"</
a>";
var shipfar = document.getElementById("shipfar");
if (shipfar) {
shipfar.innerHTML = '"' + html + '" MMSI: ' + maxmmsi + ' (Lat: '
+ maxlat + '°, Lon: ' + maxlon + '°) Range: ' + maxrange + 'nm -
Bearing: ' + maxbearing + '°';
}
maxrange = 0
});
// time out after 1hr.
if (refreshcounter++ < numrefresh)
window.setTimeout('countDown()', 1000);
else {
if (document.getElementById("countDownText"))
document.getElementById("countDownText").innerHTML = "Press
REFRESH to continue updates.";
}
}
// ==== This function displays the tooltip === It can be called
from an icon mousover or a sidebar mouseover
function showTooltip(marker) {
tooltip.innerHTML = marker.tooltip;
var point=map.getCurrentMapType().getProjection().fromLatLngToPix
el(map.fromDivPixelToLatLng(new GPoint(0,0),true),map.getZoom());
var offset=map.getCurrentMapType().getProjection().fromLatLngToPi
xel(marker.getPoint(),map.getZoom());
var anchor=marker.getIcon().iconAnchor;
var width=marker.getIcon().iconSize.width;
var height=tooltip.clientHeight;
var pos = new GControlPosition(G_ANCHOR_TOP_LEFT, new
GSize(offset.x - point.x - anchor.x + width, offset.y - point.y
-anchor.y -height));
pos.apply(tooltip);
tooltip.style.visibility="visible";
}
// === This function is invoked when the mouse goes over an entry
in the sidebar === It launches the tool tip on the icon
function mymouseover(i) {
```

```
showTooltip(gmarkers[i])
}
// === This function is invoked when the mouse leaves an entry in the sidebar === It hides the tooltip
function mymouseout() {
tooltip.style.visibility=”hidden”;
}
// Open the info box for the specified marker.
function findship(i){
gmarkers[i].openInfoWindowHtml(htmls[i]);
}
// ***SET  the data after ‘refreshInterval’ seconds. It should not be faster than ship.vbs
var refreshInterval = 120;
// ***SET numrefresh is the number of times to refresh this page before timing out.
var numrefresh = 180;
var tooltip;
var refreshcounter = 0;
var gmarkers;
var icon_k;
var icon_b;
var icon_a;
var icon_g;
var icon_r;
var icon_w;
var icon_y;
var icon_m;
var icon_o;
var icon_n;
var icon_t;
var icon_h;
var htmls;
var vectors;
var map;
var countDownTime=refreshInterval+1;
var counter;
var maxrange = 0;
var maxname, maxmmsi, maxlat, maxlon, maxbearing
function countDown(){
countDownTime--;
if (document.getElementById(“countDownText”))
document.getElementById(“countDownText”).innerHTML= “Next plot update in “ + countDownTime + “ seconds.“;
if (countDownTime<=0){
countDownTime=refreshInterval;
clearTimeout(counter)
autoUpdate()
return
}
counter=setTimeout(“countDown()”, 1000);
}
function onLoad() {
```

```
if (!GBrowserIsCompatible()) {
alert("Sorry, the Google Maps API is not compatible with this
browser");
} else {
setTimeout(loadup, 250);
}
}
//]]>
</script>
</head>
<body bgcolor="#FFFFFF" onLoad="onLoad();" onUnload="GUnload()">
<img src="http://www.kenandglen.com/images/marker.png"
style="display:none" alt="precache"/>
<img src="http://www.kenandglen.com/images/shadow50.png"
style="display:none" alt="precache"/>
<img src="http://www.kenandglen.com/images/markerTransparent.png"
style="display:none" alt="precache"/>
<img src="http://www.kenandglen.com/images/markerie.gif"
style="display:none" alt="precache"/>
<img src="http://www.kenandglen.com/images/dithshadow.gif"
style="display:none" alt="precache"/>
<img src="http://www.kenandglen.com/images/mm_20_shadow.png"
style="display:none" alt="precache"/>
<img src="http://www.kenandglen.com/images/mm_20_black.png"
style="display:none" alt="precache"/>
<img src="http://www.kenandglen.com/images/mm_20_blue.png"
style="display:none" alt="precache"/>
<img src="http://www.kenandglen.com/images/lighthouse1.png"
style="display:none" alt="precache"/>
<img src="http://www.kenandglen.com/images/mm_20_green.png"
style="display:none" alt="precache"/>
<img src="http://www.kenandglen.com/images/lifeboat2.png"
style="display:none" alt="precache"/>
<img src="http://www.kenandglen.com/images/mm_20_white.png"
style="display:none" alt="precache"/>
<img src="http://www.kenandglen.com/images/mm_20_yellow.png"
style="display:none" alt="precache"/>
<img src="http://www.kenandglen.com/images/mm_20_purple.png"
style="display:none" alt="precache"/>
<img src="http://www.kenandglen.com/images/mm_20_orange.png"
style="display:none" alt="precache"/>
<img src="http://www.kenandglen.com/images/mm_20_brown.png"
style="display:none" alt="precache"/>
<img src="http://www.kenandglen.com/images/mm_20_turkos.png"
style="display:none" alt="precache"/>
<img src="http://www.kenandglen.com/images/heli4.png"
style="display:none" alt="precache"/>
<table width="600" border="0" align="center" cellpadding="1">
<tr>
<td height="25" align="center" valign="top">My ShipPlotter Google
Map</td>
</tr>
<tr>
```

```
<td>
<div id=”map” style=”margin-left: auto; margin-right: auto;
    width: 800px; height:600px; border: s
olid #000”></div>
<div style=”text-align:center;” id=”countDownText”></div>
</td>
</tr>
<tr>
<td> <table width=”800” border=”0” align=”center”
    cellpadding=”1”>
<tr>
<td width=”100”><div align=”left”><span class=”style2”> <span
    class=”style24”>Legend:</span></span></div></td>
<td width=”99”><div align=”left”><span class=”style2”><img
    src=”images/mm_20_yellow.png” width=”12” height=”20” /> <span
    class=”style11”>Cargo </span></span></div></td>
<td width=”99”><div align=”left”><span class=”style2”><img
    src=”images/mm_20_black.png” width=”12“ height=”20” /> <span
    class=”style11”>Tanker</span></span></div></td>
<td width=”98”><div align=”left”><span class=”style2”><img
    src=”images/mm_20_purple.png” width=”12” height=”20” /> <span
    class=”style11”>Tug</span></span></div></td>
<td width=”119”><div align=”left”><span class=”style2”><img
    src=”images/mm_20_turkos.png” width=”
12” height=”20” /> <span class=”style11”>High speed craft</span></
    span></div></td>
<td width=”90”><div align=”left”><span class=”style2”><img
    src=”images/mm_20_white.png” width=”12“ height=”20” /> <span
    class=”style11”>Other</span></span></div></td>
<td width=”109” colspan=”2”><div align=”left”><span
    class=”style13”><img src=”images/lighthouse1.png” width=”12”
    height=”20” /> <span class=”style11”>Navaid </span></span></
    div></td>
</tr>
<tr style=”text-align:center;”>
<td width=”100”><div align=”left”><span class=”style2”><img
    src=”images/mm_20_blue.png” width=”12“ height=”20” /> <span class
    =”style11”>Passenger</span></span></div></td>
<td width=”99”><div align=”left”><img src=”images/
    mm_20_green.png” width=”12” height=”20” /> <span
    class=”style11”>Dredger</span></span></div></td>
<td width=”99”><div align=”left”><img src=”images/
    mm_20_brown.png” width=”12” height=”20” /> <span
    class=”style11”>Fishing</span></span></div></td>
<td width=”98”><div align=”left”><img src=”images/lifeboat2.png”
    width=”20” height=”20” /> <span class=”style11”>SAR Boat </
    span></span></div></td>
<td width=”119”><div align=”left”><img src=”images/heli4.png”
    width=”20” height=”20” /> <span class=”style11”>SAR Heli</span></
    span></div></td>
<td width=”90”><div align=”left”><span class=”style2”><img
    src=”images/mm_20_orange.png” width=”12” height=”20” /> <span
    class=”style11”>Police</span></span></div></td>
```

```
<td width="109"><div align="left"><span class="style2"><img
src="images/marker.png" width="12" height="20" /> <span
class="style11">Home Station</span></span></div></td>
</tr>
</table></td>
</tr>
<tr>
<td valign="top"><h2 class="style21">Underway</h2>
<div class="style20" id="shipmove"></div>
<h2 class="style21">Docked</h2>
<div class="style20" id="shipdocked"></div>
</td>
</tr>
</table>
</body>
</html>
```

Bibliography

Australian Maritime Safety Authority. (October 2008). Automatic Identification System (AIS) Fact Sheet. Available: http://www.amsa.gov.au/publications/fact_sheets/aisb_fact.pdf. Laast accessed September 2009.

G. E. Rogers Sr. (2009). Design your own J-Pole Antenna. Available: http://www.dxzone.com/cgi-bin/dir/jump2.cgi?ID=9284. Last accessed September 2009.

Ivor Corrigall. (2009). ShipPlotter Charts. Available: http://www.selkieman.co.uk/coggie/charts. Last accessed September 2009.

René Harte. (2009). Icom IC-M80. Available: http://discriminator.nl/ic-m80/index-en.html. Last accessed September 2009.

Satellite Signals. (2009). Latutide and Longtitude. Available: http://www.satsig.net/maps/lat-long-finder.htm. Last accessed September 2009.

Dave. (2009). Shipplotter Chart Generator. Available: http://emit.demon.co.uk/map2.php. Last accessed September 2009.

United States Coast Guard. (2009). AIS Overview. Available: http://www.navcen.uscg.gov/enav/ais/default.htm. Last accessed September 2009.

Index

A

About button 67
About ShipPlotter 53
AIS 5
AIS-2-USB 16
Alerts 60
Alert shell 50
Alert zone 50
Ambler, John 93
Antenna 16
Anti-pollution 100
API 85
API key 102
Arild Viking 76
Audio 49
Audio Input Output 62
Audio mixer 49
Autobias 40
Automatic chart selection 43
AVI 47

B

Boroysund 54
Button, Customize 24

C

Calibrate button 65, 70
Calibration 27, 41
Calshot Spit 82
Cemenmar Cuatro 4
Centre 42
Centre button 65
Centring 61
Chart 34
Chart button 65
Chart generator 25, 70
Check for new version 53
COAA Server 78
Comar 16
Config buttons 67
Container Ship 22
CSTDMA 7
Customise 43
Customize button 56

D

Data.xml 88, 90
Deegan, Curt 84, 102
Default Configuration 31
Demodulator 40
DGPS 6
Directory 49
Display calibration points 71
Dive Vessel 100
Download 43
Dredger 32, 100
DSC 7

E

ECDIS 6
Enable sharing 78
ETA 6
Exclude 59
Exit 32

F

Fairplay I viii
FAQs 97
File button 55
File Menu Commands 31
Fishing 100
Folders 21
Forum 99
FTP 47, 88

G

Garex MP-3 16
Gas Legacy 68
GMSK 8
Google.html 102
Google Earth 2, 94
Google Earth Server 64
Google Maps 2, 83
GPS 6
GPX 44
Graphic output 46
Grey ships 57

H

Halunder Jet 37
Hart, René 14
Hazard 100
HDLC 8
HDLC End Flag 40
HDLC FCS 40
Head Up 46
Help button 67
Help Menu Commands 53
Help Topics 53
High Speed Craft 37, 100
Home Centred 24
HTTP Output Server 64

I

I/O 62
I/O Monitor 37
I/O Settings 46
Image capture 46
IMO 5, 6
Input bias 40
Internet passive sharing 50

J

J-Pole 16
Jacobsson, Tore 91

L

Labels 60
LANBY 101
Lat-Long Finder 24, 70
Light Vessel 101
Load Settings 31
Local passive sharing 50
Logfile replay 50
Log File 36, 38
Log Files 64
Los Olivillos 32

M

Magnetic Bearings 58
Marine Traffic Server 81, 89, 90
Max ships 52
Mazur, Robert 92
Medical Transport 100
Messages 36
Messages button 66
Military Operations 100
MMSI 6

MoveItFreely 88
MSC Eloise 22
My ship 48

N

Nav Window 36
Nav Win button 66
Night Vision 37
NMEA 14
North up 46

O

Offshore Structure 101
Omit ships 57
Open New Chart 31
Open Street Map 2
Options Menu Commands 40
OSM 2, 46, 74
OSM button 55
OziExplorer 48

P

Photoshop 25
Pilot Vessel 100
Port Tenders 100
Preamble 40
PreAmp 16
Predicted track 56
Predict positions 58
Preemphasis 40
Process Menu Commands 37
Purge 52

Q

Queen Mary 2 96
Quick chart 43
Quick Chart button 66

R

RACON 101
Radar 33, 46
Radar button 65
Range rings 58
raster chart 42
Raw button 66
Registering ShipPlotter 53
RORO 101
ROT 58
Route 36, 49
Route button 66

S

Salvamar Algeciras 18
Satellite 43
Satellite button 55
Save calibration data 71
Save Settings 31
Scale 61
Scheduling 89
Scrap ships 58
Search And Rescue 18, 100
Serial File 38
Serial Input 63
Server 50
Share button 65, 79
Sharing 49, 77
Shell 52
Shell button 67
Ship9.vbs 85, 86
Shipinfo.txt 32
Shipnames.txt 35
Ships 35
Ships button 65
Ship Info 36
Signal 36
Siri Knutsen 12
SOLAS 5
SOTDMA 7, 9
Soundcard 49
SP2GM.vbs 83
Special craft 100
SPTools 85, 88
SR162 15
Start 38
Start button 55
Status 59
Status Bar 33
Steam Ship 54
Stop 39
Stop button 55

T

Tanker 12, 68, 100
TCP/IP peer-to-peer 63
TDMA 7
Text colours 65
The World 39
Toolbar 33, 55
Towing 100
Trail 58
Tug 76, 100, viii
Type 59

U

UDP/IP peer-to-peer 63

V

Vb4ship 102
Vector button 55
Vector chart 42
Vector map 75
Vmap 44
VTS 101

W

Wake history 57
Wave button 55
Wave File 38
Wing In Ground 100

Z

Zoom buttons 65, 70

1106934R0

Printed in Great Britain by
Amazon.co.uk, Ltd.,
Marston Gate.